Diophantine Approximation and Its Applications

ACADEMIC PRESS RAPID MANUSCRIPT REPRODUCTION

Proceedings of a Conference on
Diophantine Approximation and Its Applications
Held in Washington D.C., June 6-8, 1972

Conference on Diophantine Approximation and Its Applications, Washington, D.C., 1972.

Diophantine Approximation and Its Applications

Edited by

Charles F. Osgood

Code 7840
Naval Research Laboratory
Washington, D. C.

ACADEMIC PRESS
New York and London 1973
A Subsidiary of Harcourt Brace Jovanovich, Publishers

ACADEMIC PRESS, INC.
111 Fifth Avenue, New York, New York 10003

United Kingdom Edition published by
ACADEMIC PRESS, INC. (LONDON) LTD.
24/28 Oval Road, London NW1

Library of Congress Cataloging in Publication Data

Conference on Diophantine Approximation and Its Applications, Washington, D. C., 1972.
Diophantine approximation and its applications.

1. Diophantine analysis–Congresses.
2. Approximation theory–Congresses. I. Osgood, Charles F., ed. II. Title.
QA242.C65 1972 512'.74 72-7696
ISBN 0–12–528650–3

PRINTED IN THE UNITED STATES OF AMERICA

CONTENTS

CONTRIBUTORS

A. Baker
Trinity College, Cambridge, England

R. T. Bumby
Rutgers College, Department of Mathematics, New Brunswick, New Jersey 08903

P. L. Cijsouw
Instituut voor Propedeutische Wiskunde, University of Amsterdam, Amsterdam, The Netherlands

John Coates
The Department of Mathematics, Harvard University, Cambridge, Massachusetts 02138

Janos Galambos
The Department of Mathematics, Temple University, Philadelphia, Pennsylvania 19122

Carlos J. Moreno
The Department of Mathematics, The University of Illinois, Urbana, Illinois 61801

H. Niederreiter
The Department of Mathematics, Southern Illinois University, Carbondale, Illinois 62901

Charles F. Osgood
Mathematics Research Center–Code 7840, Naval Research Laborato·
Washington, D. C. 20375

Walter Philipp
The Department of Mathematics, The University of Illinois, Urbana, Illinois 61801

Wolfgang M. Schmidt
The Department of Mathematics, The University of Colorado, Boulder, Colorado 80302

H. M. Stark
The Department of Mathematics, M.I.T., Cambridge, Massachusetts 02139

E. G. Straus
The Department of Mathematics, U.C.L.A., Los Angeles, California 90024

R. Tijdeman
Mathematisch Instituut der Rijksuniversiteit Leiden, Leiden, The Netherlands

Paul Turán
Math. Inst. Univ., Muzeum krt. 6-8, Budapest VIII, Hungary

S. K. Zaremba
Centre de Recherches Mathématiques, Université de Montréal, 90 Bellingham, E-525, Montreal, Canada

PREFACE

This volume represents the Proceedings of a Conference on Diophantine Approximation and Its Applications held in Washington, D. C., June 6-8, 1972, and sponsored by the Mathematics Research Center of the Naval Research Laboratory.

The purpose of this meeting was to stimulate research in the area of diophantine approximation by bringing together many of the leading researchers in this field so that they could exchange information and ideas.

Approximately 40 mathematicians registered for the conference which consisted of 14 formal lectures, one on each of the papers contained in this volume.

Many thanks go to all of the authors whose papers appear here for the necessary cooperation to produce this volume so relatively soon after the conference. My thanks go to Dr. Paul B. Richards, Superintendent of the Mathematics and Information Sciences Division of NRL, who suggested this conference and gave it his complete support.

Finally, my thanks go to all those other members of the Mathematics and Information Sciences Division of NRL whose valuable aid helped to make this conference a success.

Diophantine Approximation and Its Applications

A CENTRAL THEOREM IN TRANSCENDENCE THEORY

A. Baker

1. Introduction.

Much research has been carried out in the last few years in connection with the theory of linear forms in the logarithms of algebraic numbers, and I should like to discuss here a new theorem in this field which includes many of the more recent results, as well as most of the classical propositions in the subject, as special cases.

Let $\alpha_1,\ldots,\alpha_n$ denote non-zero algebraic numbers with degrees at most d and let the heights of $\alpha_1,\ldots,\alpha_{n-1}$ and α_n be at most A' and $A(\geq 4)$ respectively. Further let $\beta_0,\ldots,\beta_n$ denote algebraic numbers with degrees and heights at most d and $B(\geq 4)$ respectively. We put

$$\Lambda = \beta_0 + \beta_1 \log \alpha_1 + \ldots + \beta_n \log \alpha_n$$

for some fixed determinations of the logarithms, and prove:

THEOREM. If $\Lambda \neq 0$ then

$$|\Lambda| > (B \log A)^{-C \log A},$$

where C is effectively computable in terms of A', n and d only. †

It will be recalled (see[1, III]) that $\Lambda \neq 0$ is implied by the linear independence of $\log \alpha_1, \ldots, \log \alpha_n$ over the rationals, assuming that $\beta_0, \ldots, \beta_n$ are not all 0, and the latter could therefore be adopted as the hypothesis of the theorem. The result is plainly best possible with respect to B when A is fixed and, except for the bracketed factor log A, also with respect to A when B is fixed. Furthermore, in addition to incorporating, as mentioned above, several earlier theorems in this context (cf. [1,2,5-10]), it also furnishes some new results in classical transcendence theory. For instance, it implies that if $1, \beta_1, \ldots, \beta_n$ are linearly independent over the rationals then the estimate

$$|\alpha_1^{\beta_1} \ldots \alpha_n^{\beta_n} - \gamma| > H^{-C \log \log H}$$

is valid for all algebraic numbers γ with height at most $H (\geqslant 4)$, where $C > 0$ depends only on the α's, β's and the degree of γ. In particular, the case $n = 1$ shows that

$$|e^{\pi} - p/q| > q^{-c \log \log q}$$

for all rationals $p/q (q \geqslant 4)$, where c is an absolute constant; and this is the best measure of irrationality for e^{π} established to date.

Although, in general, it seems difficult to eliminate the bracketed factor log A from the bound for $|\Lambda|$, this

† The value of C depends also on the original determinations of the logarithms.

has in fact been achieved in some particular cases. Indeed when $\beta_o = 0$ and $\beta_1,\ldots,\beta_n$ are rational integers, it has been shown that

$$|\Lambda| > C^{-\log A \log B},$$

and moreover, under the further specialization $\beta_n = -1$, we have

$$|\Lambda| > C^{-\log A} e^{-\varepsilon B}$$

for any $\varepsilon > 0$, where now $C > 0$ depends only on A', n, d and ε [3,4]. The latter result is fundamental to the theory of rational approximations to algebraic numbers and will probably find application in several other spheres. The arguments used in the demonstration are similar to those employed here, but the underlying auxiliary function is substantially modified and the logical structure of the argument is consequently more complicated.

2. Preliminaries.

For any integers $k \geqslant 1$, $\ell \geqslant 0$ we signify by $\nu(\ell,k)$ the least common multiple of $\ell + 1,\ldots,\ell + k$. We define

$$\Delta(x;k) = (x + 1)\ldots(x + k)/k!\ ,$$

and we put

$$\Delta(x;k,\ell,m) = \frac{1}{m!}\frac{d^m}{dx^m}(\Delta(x;k))^\ell.$$

We shall require the following lemmas.

LEMMA 1. $(\nu(x,k))^m\Delta(x;k,\ell,m)$ is a positive integer when x is a positive integer and we have

$$\Delta(x;k,\ell,m) \leqslant 4^{\ell(x+k)}, \quad \nu(x,k) \leqslant \{c(x+k)/k\}^{2k}$$

for some absolute constant c.

Proof. This is Lemma 1 of [3].

LEMMA 2. If $P(x)$ is a polynomial with degree $n > 0$ and with coefficients in a field K then, for any integer m with $0 \leqslant m \leqslant n$, the polynomials $P(x), P(x+1),\ldots,P(x+m)$ and $1, x,\ldots,x^{n-m-1}$ are linearly independent over K.

Proof. This is Lemma 2 of [3].

LEMMA 3. Let $\alpha_1,\ldots,\alpha_n$ be nonzero elements of an algebraic number field K and let $\alpha_1^{1/p},\ldots,\alpha_n^{1/p}$ denote fixed pth roots for some prime p. Further let $K' = K(\alpha_1^{1/p},\ldots,\alpha_{n-1}^{1/p})$. Then either $K'(\alpha_n^{1/p})$ is an extension of K' of degree p or we have

$$\alpha_n = \alpha_1^{j_1}\ldots\alpha_{n-1}^{j_{n-1}}\gamma^p$$

for some γ in K and some integers $j_1,\ldots,j_{n-1}$ with $0 \leqslant j_r < p$.

Proof. This is Lemma 3 of [5].

LEMMA 4. Suppose that α,β are elements of an algebraic number field with degree D and that for some positive integer p we have $\alpha = \beta^p$. If $a\alpha$ is an algebraic integer for some positive rational integer a and if b is the leading coefficient in the minimal defining polynomial of β then $b \leqslant a^{D/p}$.

Proof. This is Lemma 4 of [5].

3. Main Lemmas.

We denote by $\alpha_1,\ldots,\alpha_n$, where $n \geqslant 2$, nonzero algebraic numbers with degrees at most d, and we suppose that the heights of $\alpha_1,\ldots,\alpha_{n-1}$ and α_n do not exceed A' and $A(\geqslant 4)$ respectively. By $\log \alpha_1,\ldots,\log \alpha_n$ we signify any fixed determinations of the logarithms, and by c, c_1, $c_2,\ldots$ numbers, greater than 1, that depend only on n, d, A' and these determinations. We suppose that there exist algebraic numbers $\beta_o,\ldots,\beta_{n-1}$ with degrees at most d and heights at most $B(\geqslant 4)$ such that

$$0 < |\beta_o + \beta_1 \log \alpha_1 + \ldots + \beta_{n-1} \log \alpha_{n-1} - \log \alpha_n| < (B \log A)^{-C \log A}, \tag{1}$$

where $C = C(n,d,A')$ is assumed sufficiently large for the validity of the subsequent arguments. We shall proceed to show that there exist then further algebraic numbers $\beta'_o,\ldots,\beta'_{n-1}$ with degrees at most d^{2n} and heights at most $c_1 B$, and an algebraic number α'_n in the field generated by the α's over the rationals with height at most $c_2 A^{1/2}$, such that (1) holds with $\beta_o,\ldots,\beta_{n-1}$ and α_n replaced by $\beta'_o,\ldots,\beta'_{n-1}$ and α'_n respectively; an inductive argument will then complete the proof of the theorem.

We signify by k an integer exceeding a sufficiently large number c as above, and we write

$$h = [\log(k\, B \log A)], \qquad L_{-1} = h - 1 ,$$

$$L = L_o = \ldots = L_{n-1} = [k^{1-1/(4n)} \log A], \quad L_n = [k^{1/2}] ,$$

where, as usual, $[x]$ denotes the integral part of x. Further we write

$$f_{m_0,\ldots,m_{n-1}}(z_0,\ldots,z_{n-1})$$

$$= (\partial/\partial z_0)^{m_0}\ldots(\partial/\partial z_{n-1})^{m_{n-1}} f(z_0,\ldots,z_{n-1}) .$$

LEMMA 5. There are integers $p(\lambda_{-1},\ldots,\lambda_n)$, not all 0, with absolute values at most $A^{c_3 hk}$, such that the function

$$\Phi(z_0,\ldots,z_{n-1}) = \sum_{\lambda_{-1}=0}^{L_{-1}} \cdots \sum_{\lambda_n=0}^{L_n} p(\lambda_{-1},\ldots,\lambda_n)$$

$$(\Delta(z_0+\lambda_{-1};h))^{\lambda_0+1} e^{\lambda_n \beta_0 z_0} \alpha_1^{\gamma_1 z_1} \ldots \alpha_{n-1}^{\gamma_{n-1} z_{n-1}} ,$$

where $\gamma_r = \lambda_r + \lambda_n \beta_r (1 \leqslant r < n)$, satisfies

$$(2) \qquad |\Phi_{m_0,\ldots,m_{n-1}}(\ell,\ldots,\ell)| < (B \log A)^{-\frac{1}{2} C \log A}$$

for all integers ℓ with $1 \leqslant \ell \leqslant h$ and all nonnegative integers $m_0,\ldots,m_{n-1}$ with $m_0+\ldots+m_{n-1} \leqslant k \log A$.

Proof. We determine the $p(\lambda_{-1},\ldots,\lambda_n)$ such that

$$(3) \qquad \sum_{\lambda_{-1}=0}^{L_{-1}} \cdots \sum_{\lambda_n=0}^{L_n} p(\lambda_{-1},\ldots,\lambda_n) q(\lambda_{-1},\lambda_0,\lambda_n,\ell) \alpha_1^{\lambda_1 \ell} \ldots \alpha_n^{\lambda_n \ell} \gamma_1^{m_1} \ldots \gamma_{n-1}^{m_{n-1}} = 0$$

for the above ranges of ℓ and $m_o,\ldots,m_{n-1}$, where

$$q(\lambda_{-1},\lambda_o,\lambda_n,z) = \sum_{\mu_o=0}^{m_o} \binom{m_o}{\mu_o} \mu_o!$$

$$\Delta(z+\lambda_{-1};\ h,\ \lambda_o+1,\ \mu_o)(\lambda_n\beta_o)^{m_o-\mu_o}\ .$$

We shall verify subsequently that (3) implies (2). Let $a_1,\ldots,a_n,\ b_o,\ldots,b_{n-1}$ denote the leading coefficients (supposed positive) in the minimal defining polynomials of $\alpha_1,\ldots,\alpha_n,\ \beta_o,\ldots,\beta_{n-1}$ respectively. For any nonnegative integer j we have

$$(a_r\alpha_r)^j = \sum_{s=0}^{d-1} a_{rs}^{(j)}\alpha_r^s, \quad (b_r\beta_r)^j = \sum_{t=0}^{d-1} b_{rt}^{(j)}\ \beta_r^t\ ,$$

where the $a_{rs}^{(j)}$, $b_{rt}^{(j)}$ denote rational integers; the $a_{rs}^{(j)}$ have absolute values at most c_4^j or $(2A)^j$ according as $r < n$ or $r = n$, and the $b_{rt}^{(j)}$ have absolute values at most $(2B)^j$. Thus, on multiplying (3) by

$$P = a_1^{L_1\ell}\ldots a_n^{L_n\ell}\, b_o^{m_o}\ldots b_{n-1}^{m_{n-1}}\ ,$$

writing

$$\gamma_r^{m_r} = \sum_{\mu_r=0}^{m_r}\binom{m_r}{\mu_r}\lambda_r^{m_r-\mu_r}(\lambda_n\beta_r)^{\mu_r}\ ,$$

and substituting for the powers of $a_r\alpha_r$ and $b_r\beta_r$ which result, we obtain

$$\sum_{s_1=0}^{d-1}\cdots\sum_{s_n=0}^{d-1}\ \sum_{t_o=0}^{d-1}\cdots\sum_{t_{n-1}=0}^{d-1} V(s,t)\alpha_1^{s_1}\ldots\alpha_n^{s_n}\,\beta_o^{t_o}\ldots\beta_{n-1}^{t_{n-1}}=0,$$

where

$$V(s,t) = \sum_{\lambda_{-1}=0}^{L_{-1}} \cdots \sum_{\lambda_n=0}^{L_n} \sum_{\mu_o=0}^{m_o} \cdots \sum_{\mu_{n-1}=0}^{m_{n-1}} p(\lambda_{-1},\ldots,\lambda_n) q' q'' q'''$$

and q', q'', q''' are given by

$$q' = \prod_{r=1}^{n} \left\{ a_r^{(L-\lambda_r)\ell} \, a_{r,s_r}^{(\lambda_r \ell)} \right\},$$

$$q'' = \prod_{r=1}^{n-1} \left\{ \binom{m_r}{\mu_r} (b_r \lambda_r)^{m_r - \mu_r} \lambda_n^{\mu_r} \, b_{r,t_r}^{(\mu_r)} \right\},$$

$$q''' = \binom{m_o}{\mu_o} \mu_o! \; \Delta(z+\lambda_{-1};\, h,\, \lambda_o+1,\, \mu_o) \lambda_n^{m_o-\mu_o} \, b_n^{\mu_o} \, b_{o,t_o}^{(m_o-\mu_o)} .$$

Thus (3) will be satisfied if the d^{2n} equations $V(s,t) = 0$ hold. Now these represent $M \leqslant d^{2n} h(k \log A+1)^n$ linear equations in the $N = (L_{-1} + 1)\ldots(L_n + 1)$ unknowns $p(\lambda_{-1},\ldots,\lambda_n)$. Further, Lemma 1 shows that, after multiplying by $(\nu(0,\, 3h))^{m_o}$, the coefficients in these equations will be rational integers. Furthermore, we have

$$\Delta(\ell + \lambda_{-1};\, h,\, \lambda_o + 1,\, \mu_o) \leqslant c_5^{Lh}$$

and since $kB \log A \leqslant e^{h+1}$, we see that

$$|q'| \leqslant c_6^{Lh} (2A)^{L_n h}, \quad |q''| \leqslant e^{2h(m_1+\ldots+m_{n-1})},$$

$$|q'''| \leqslant 2^{m_o} (\mu_o b_n)^{\mu_o} (2B\lambda_n)^{m_o-\mu_o} c_5^{Lh} \leqslant e^{2hm_o} c_5^{Lh} .$$

Since also $\nu(0,\, 3h) \leqslant c_7^h$, it follows that the coefficients have absolute values at most $U = A^{c_8 hk}$. Now

$$N > hk^{n+1/4}(\log A)^n > 2M$$

and hence the system of equations $V(s,t) = 0$ can be solved nontrivially and the integers $p(\lambda_{-1},\ldots,\lambda_n)$ can be chosen to have absolute values at most $NU \leq A^{c_3 hk}$.

It remains only to verify that (3) implies (2). Now the left-hand side of (3) is obtained from $\Phi_{m_0,\ldots,m_{n-1}}(\ell,\ldots,\ell)$, apart from a factor

$$P' = (\log \alpha_1)^{m_1} \ldots (\log \alpha_{n-1})^{m_{n-1}}$$

in the latter, by substituting α_n for $\alpha_n' = e^{\beta_0}\alpha_1^{\beta_1}\ldots\alpha_{n-1}^{\beta_{n-1}}$. From (1) we have

$$|\log \alpha_n' - \log \alpha_n| < (B \log A)^{-C \log A}$$

for some value of the first logarithm, and since, for any complex number z,

$$|e^z - 1| \leq |z| e^{|z|},$$

and moreover $|\alpha_n| \leq dA$, we obtain

$$(4) \qquad |\alpha_n' - \alpha_n| < (B \log A)^{-\frac{3}{4} C \log A}.$$

Also we have

$$|\alpha_n'^{\lambda_n \ell} - \alpha_n^{\lambda_n \ell}| \leq (4\, dA)^{L\ell} |\alpha_n' - \alpha_n| ,$$

and estimates similar to those employed above show that

$$|P'| \leq c_9^{k \log A} , \quad |q(\lambda_{-1},\lambda_0,\lambda_n,\ell)| \leq c_{10}^{(L+m_0)h} ,$$

$$|\gamma_r| \leqslant e^{2h}, \quad |\alpha_r^{\lambda_r \ell}| \leqslant c_{11}^{Lh} \ (1 \leqslant r < n) .$$

Thus we see that

$$|\Phi_{m_o,\ldots,m_{n-1}}(\ell,\ldots,\ell)| \leqslant NA^{c_{12}hk} |\alpha'_n - \alpha_n| .$$

But clearly $N \leqslant e^{2hn}$ and

$$h \leqslant \log(B \log A) + \log k ,$$

and hence (2) follows if $C > c_{13} k \log k$.

LEMMA 6. For any nonnegative integers $m_o,\ldots,m_{n-1}$ with $m_o+\ldots+m_{n-1} \leqslant k \log A$, we have

(5) $$|\Phi_{m_o,\ldots,m_{n-1}}(z,\ldots,z)| \leqslant A^{c_{14}hk} c_{15}^{L|z|} .$$

Further, for any integer ℓ with $h < \ell \leqslant hk^{2n}$, either (2) holds or

(6) $$|\Phi_{m_o,\ldots,m_{n-1}}(\ell,\ldots,\ell)| \geqslant A^{-c_{16}hk\{1+\log(\ell/h)\}} c_{17}^{-L\ell} .$$

Proof. We have

$$\Phi_{m_o,\ldots,m_{n-1}}(z,\ldots,z) = P' \sum_{\lambda_{-1}=0}^{L_{-1}} \cdots \sum_{\lambda_n=0}^{L_n} p(\lambda_{-1},\ldots,\lambda_n)$$

$$q(\lambda_{-1},\lambda_o,\lambda_n,z) e^{\lambda_n \beta_o z} \alpha_1^{\gamma_1 z} \cdots \alpha_{n-1}^{\gamma_{n-1} z} \gamma_1^{m_1} \cdots \gamma_{n-1}^{m_{n-1}} ,$$

where P' and $q(\lambda_{-1},\lambda_o,\lambda_n,z)$ are defined as in Lemma 5. Now (4) implies that

$$|\alpha_n'^{z}| \leqslant e^{(|\log \alpha_n|+1)|z|} \leqslant A^{c_{18}|z|} ,$$

and clearly

$$|\alpha_1^{\lambda_1 z} \ldots \alpha_{n-1}^{\lambda_{n-1} z}| \leqslant c_{19}^{L|z|} .$$

Furthermore, since

$$|z + \lambda_{-1}| \leqslant [|z|] + h ,$$

we deduce from Lemma 1 that

$$|\Delta(z+\lambda_{-1}; h, \lambda_o+1, \mu_o)| \leqslant c_{20}^{L(|z|+h)} .$$

This gives

$$|q(\lambda_{-1}, \lambda_o, \lambda_n, z)| \leqslant e^{2hm_o} c_{20}^{L(|z|+h)} ,$$

and the required estimate (5) now follows easily as in the latter part of the proof of Lemma 5.

To prove the second assertion we begin by noting that the expression on the left of (2), say Q, is an algebraic number with degree at most d^{2n}. Further, by estimates similar to those given above, it is readily verified that each conjugate of Q, obtained by substituting arbitrary conjugates for $\alpha_1,\ldots,\alpha_n$, has absolute value at most $A^{c_{21}hk} c_{22}^{L\ell}$. Furthermore, from Lemma 1, we see that, on multiplying Q by

$$(\nu(\ell,2h))^{m_o} P \leqslant (c_{23}\ell/h)^{4hm_o} c_{24}^{L\ell} ,$$

we obtain an algebraic integer. Hence we conclude that either $Q = 0$ or

$$|Q| \geqslant A^{-c_{25}hk} c_{26}^{-L\ell} (\ell/h)^{-c_{27}hm_o} ;$$

and since $m_o \leqslant k \log A$, the number on the right exceeds the right hand side of (6) for some c_{16} and c_{17}. But, as above, we deduce easily from (4) that the number

$$P'^{-1}\Phi = P'^{-1}\Phi_{m_o,\ldots,m_{n-1}}^{(\ell,\ldots,\ell)}$$

differs from Q by at most an amount

$$A^{c_{28}\ell k - \frac{3}{4} C \log(B \log A)} .$$

If $\ell \leqslant hk^{2n}$ and $C > c_{29}k^{2n+1}(\log k)^2$, this is at most $\frac{1}{2}|Q|$, and hence, if $Q \neq 0$, we obtain $|\Phi| > \frac{1}{2} P'|Q|$, which gives (6).

<u>LEMMA 7</u>. Suppose that $0 < \varepsilon < (2^8 n\, c_{16})^{-1}$ where c_{16} is the number indicated in (6). Then, for any integer J with $0 \leqslant J < 2n/\varepsilon$, (2) is satisfied for all integers ℓ with $1 \leqslant \ell \leqslant hk^{\varepsilon J}$ and all nonnegative integers $m_o,\ldots,m_{n-1}$ with

$$m_o + \ldots + m_{n-1} \leqslant (k/2^J)\log A .$$

<u>Proof</u>. The lemma holds for $J = 0$ by Lemma 5. We suppose that K is an integer satisfying $0 \leqslant K \leqslant (2n/\varepsilon) -1$ and we assume that the lemma has been verified for $J = 0, 1,\ldots,K$. We proceed to prove the lemma for $J = K+1$.

We begin by defining

$$R_J = [hk^{\varepsilon J}], \quad S_J = [(k/2^J)\log A] \quad (J = 0,1,\ldots) .$$

It suffices then to prove that for any integer ℓ with $R_K < \ell \leqslant R_{K+1}$ and any set of nonnegative integers $m_o,\ldots,m_{n-1}$ with $m_o+\ldots m_{n-1} \leqslant S_{K+1}$ we have

$$(7) \qquad |f(\ell)| < (B \log A)^{-\frac{1}{2} C \log A},$$

where

$$f(z) = \Phi_{m_o,\ldots,m_{n-1}}(z,\ldots,z) .$$

Since clearly $f_m(r)$ is given by

$$(\partial/\partial z_o + \ldots + \partial/\partial z_{n-1})^m \Phi_{m_o,\ldots,m_{n-1}}(z_o,\ldots,z_{n-1})$$

evaluated at $z_o = \ldots = z_{n-1} = r$, we have, by our inductive hypothesis,

$$(8) \qquad |f_m(r)| < n^m(B \log A)^{-\frac{1}{2} C \log A} (1 \leqslant r \leqslant R_K,$$

$$0 \leqslant m \leqslant S_{K+1}) .$$

We write, for brevity,

$$F(z) = \{(z-1) \ldots (z-R_K)\}^{S_{K+1}+1}$$

and we denote by Γ the circle in the complex plane, described in the positive sense, with center the origin and radius $R = R_{K+1}k^{1/(8n)}$. By Cauchy's residue theorem we have

$$(9) \qquad \frac{1}{2\pi i}\int_\Gamma \frac{f(z)dz}{(z-\ell)F(z)} = \frac{f(\ell)}{F(\ell)} + \frac{1}{2\pi i}\sum_{r=1}^{R_K}\sum_{m=0}^{S_{K+1}} \frac{f_m(r)}{m!} \int_{\Gamma_r} \frac{(z-r)^m dz}{(z-\ell)F(z)} ,$$

where Γ_r denotes the circle in the complex plane, described in the positive sense, with center r and radius $\frac{1}{2}$. Since, for z on Γ_r,

$$|(z-r)^m/F(z)| \leqslant \{\tfrac{1}{8}(R_K-r-1)!(r-2)!\}^{-S_{K+1}-1}$$

$$\leqslant 8^{R_K S_{K+1}}(R_K!)^{-S_{K+1}-1},$$

we deduce from (8) that the absolute value of the double sum on the right of (9) is at most

$$R_K(S_{K+1}+1)8^{R_K S_{K+1}+1}(R_K!)^{-S_{K+1}-1}(B\log A)^{-\frac{1}{4}C\log A}.$$

Further, for $R_K < \ell \leqslant R_{K+1}$, we have

$$|F(\ell)| = \{(\ell-1)!/(\ell-R_K-1)!\}^{S_{K+1}+1} \leqslant (2^{R_{K+1}}R_K!)^{S_{K+1}+1},$$

and, since $R_{K+1} \leqslant hk^{2n}$, we see that if (6) holds then

$$|f(\ell)| > (B\log A)^{-\frac{1}{8}C\log A}$$

whence the absolute value of the number on the right of (9) exceeds $\frac{1}{2}|f(\ell)/F(\ell)|$. We proceed to show that the assumption that (6) is valie leads to a contradiction.

Let θ and Θ denote respectively the upper bound of $|f(z)|$ and the lower bound of $|F(z)|$ with z on Γ. Since $2|z-\ell|$ with z on Γ exceeds the radius of Γ, we obtain from (9)

(10) $$4\,\theta|F(\ell)| > \Theta|f(\ell)|.$$

Now clearly we have $\Theta \geqslant (\frac{1}{2}R)^{R_K(S_{K+1}+1)}$ and thus

$$(11)\qquad \log(\Theta|F(\ell)|^{-1}) \geqslant R_K(S_{K+1}+1)\log(\tfrac{1}{2}k^{1/(8n)}) .$$

Further, from (5) we see that $\theta \leqslant A^{c_{14}hk}\, c_{15}^{LR}$ and so, by virtue of (6),

$$(12)\qquad \log(\theta|f(\ell)|^{-1}) \leqslant \{c_{14} + 2c_{16}\log(R_{K+1}/h)\}$$

$$hk \log A + c_{28}LR .$$

But the number on the right of (11) is at least

$$2^{-K-6}\, n^{-1}\, hk^{\varepsilon K+1} \log k \log A ,$$

and that on the right of (12) is at most

$$\{c_{14} + 2c_{16}\, \varepsilon(K+1)\log k + c_{28}k^{\varepsilon(K+1)-1/(8n)}\}hk \log A .$$

Further, with the bound for ε specified in the lemma, we have $2c_{16}\varepsilon < 2^{-7}n^{-1}$ and $\varepsilon < (8n)^{-1}$. Thus (10) is untenable if k is sufficiently large; the contradiction implies the validity of (2) and the lemma follows by induction.

<u>LEMMA 8</u>. For all integers $\ell, m_o,\ldots,m_{n-1}, q$ with

$$1 \leqslant \ell \leqslant hk,\ 0 \leqslant m_r \leqslant L(0\leqslant r<n),\ 2 < q \leqslant 2L_n,(\ell,q) = 1 ,$$

(3) is satisfied with ℓ replaced by ℓ/q.

<u>Proof</u>. From Lemma 7 we see that (2) holds for all integers ℓ with $1 \leqslant \ell \leqslant X$ and all nonnegative integers $m_o,\ldots,m_{n-1}$ with $m_o+\ldots+m_{n-1} \leqslant 2Y$, where

$$X = [hk^{n+1}],\qquad Y = [c_{29}^{-1}\, k \log A]$$

and $c_{29} = 2^{(2n/\varepsilon)+1}$. Further, as in the proof of Lemma 7, it follows that $f(z) = \Phi_{m_0,\ldots,m_{n-1}}(z,\ldots,z)$, where $0 \leqslant m_r \leqslant L$, satisfies

$$|f_m(r)| < n^m(B \log A)^{-\frac{1}{2} C \log A} \tag{13}$$

$$(1 \leqslant r \leqslant X, \quad 0 \leqslant m \leqslant Y) .$$

On writing for brevity,

$$E(z) = \{(z-1) \ldots (z-X)\}^{Y+1}$$

and denoting by Γ the circle in the complex plane, described in the positive sense, with center the origin and radius $R = Xk^{1/(8n)}$, we deduce from Cauchy's residue theorem

$$\frac{1}{2\pi i}\int_{\Gamma} \frac{f(z)dz}{(z-\ell/q)E(z)} = \frac{f(\ell/q)}{E(\ell/q)}$$

$$+ \frac{1}{2\pi i}\sum_{r=1}^{X}\sum_{m=0}^{Y}\frac{f_m(r)}{m!}\int_{\Gamma_r}\frac{(z-r)^m dz}{(z-\ell/q)E(z)} ,$$

where ℓ, q are any integers satisfying the hypotheses of the lemma and Γ_r denotes the circle in the complex plane described in the positive sense with center r and radius $1/(2q)$. Since for z on Γ_r,

$$|(z-r)^m/E(z)| \leqslant \{(8q)^{-1}(X-r-1)!(r-2)!\}^{-Y-1}$$

$$\leqslant 8^{XY}(8q)^{Y+1}(X!)^{-Y-1} ,$$

it follows from (13) that the absolute value of the double sum on the right of the above equation is at most

$$X(Y+1)8^{XY}(8q)^{Y+2}(X!)^{-Y-1}(B \log A)^{-\frac{1}{4} C \log A} .$$

Further, by virtue of Lemma 6, we have, for any z on Γ,

$$|f(z)| < A^{c_{14}hk} c_{15}^{LR} ,$$

and it is clear that

$$|E(z)| \geqslant (\tfrac{1}{2} R)^{X(Y+1)} ,$$

$$|E(\ell/q)| \leqslant (2X)^{X(Y+1)} \leqslant 8^{X(Y+1)}(X!)^{Y+1} .$$

Thus we obtain

$$|f(\ell/q)| < A^{c_{14}hk} c_{15}^{LR}(\tfrac{1}{4} k^{1/(8n)})^{-X(Y+1)} + (B \log A)^{-\frac{1}{8}C \log A} ,$$

and since $Lk^{1/(8n)} < k$ and

$$c_{30}^{-1} hk^{n+2} \log A < X(Y+1) < \frac{1}{16} C \log A \log(B \log A) ,$$

the number on the right is at most e^{-XY}.

We now utilize the latter estimate to confirm the validity of (3) with ℓ replaced by ℓ/q. Let the left-hand side of (3) thus modified be denoted by Q. Clearly Q is an algebraic number with degree at most $d^{2n} q^{n}$, and each conjugate has absolute value at most $A^{c_{31}hk^2}$. Further, it is easily verified from the identity

$$\Delta(x;\, k,\, \ell,\, m) = (\Delta(x;k))^{\ell} \sum((x+j_1)\ldots(x+j_m))^{-1} ,$$

where the summation is over all selections of m integers $j_1,\ldots,j_m$ from $1,\ldots,k$ repeated ℓ times, and the right-hand side is read as 0 if $m > k\ell$, that on multiplying Q by

$$P_q^{4L_o h}(\nu(0,2hk))^{m_o} \leqslant A^{c_{32}hk^2} ,$$

where P is given as in Lemma 5, one obtains an algebraic integer. Thus, if $Q \neq 0$, we have

$$|Q| \geqslant A^{-c_{33}hk^2 q^n} .$$

But is is easily seen from (4) that

$$|Q - P'^{-1}f(\ell/q)| < (B \log A)^{-\frac{1}{2} C \log A} ,$$

whence $|f(\ell/q)| \geqslant \frac{1}{2}|P'Q|$. Since $q \leqslant 2k^{\frac{1}{2}}$, it is clear that the estimate for $|Q|$ given above is inconsistent with the upper bound e^{-XY} for $|f(\ell/q)|$ obtained earlier. Hence we conclude that $Q = 0$, as required.

4. Proof of the Theorem.

It is well-known that there exists at least one prime p between L_n and $2L_n$ exclusive and we take $q = p$ in Lemma 8. On writing (3) with ℓ replaced by ℓ/p in the form

$$\sum_{\lambda_n=0}^{L_n} \left(\sum_{\lambda_{-1}=0}^{L_{-1}} \cdots \sum_{\lambda_{n-1}=0}^{L_{n-1}} p(\lambda_{-1},\ldots,\lambda_n) q(\lambda_{-1},\lambda_o,\lambda_n,\ell/p) \right.$$

$$\left. \gamma_1^{m_1} \cdots \gamma_{n-1}^{m_{n-1}} \alpha_1^{\lambda_1 \ell/p} \cdots \alpha_{n-1}^{\lambda_{n-1}\ell/p} \right) \alpha_n^{\lambda_n \ell/p} = 0 ,$$

we see that Lemma 3 implies that either each of the expressions in parenthesis is 0 for all $m_o,\ldots,m_{n-1}$ with

$0 \leqslant m_r \leqslant L$ or

$$\alpha_n^{\ell} = \alpha_1^{\ell j_1} \cdots \alpha_{n-1}^{\ell j_{n-1}} \alpha'^{p} \tag{14}$$

for some integers $j_1, \ldots, j_{n-1}$ with $0 \leqslant j_r < p$ and some element α' in the field K generated by the α's and β's over the rationals. We first show that the above expressions cannot all vanish for every ℓ with $1 \leqslant \ell \leqslant hk$ and $(\ell, p) = 1$.

In fact the latter would imply that

$$\sum_{\lambda_{n-1}=0}^{L_{n-1}} \left(\sum_{\lambda_{-1}=0}^{L_{-1}} \cdots \sum_{\lambda_{n-2}=0}^{L_{n-2}} p(\lambda_{-1}, \ldots, \lambda_n) q(\lambda_{-1}, \lambda_o, \lambda_n, \ell/p) \right.$$

$$\left. \gamma_1^{m_1} \cdots \gamma_{n-2}^{m_{n-2}} \alpha_1^{\lambda_1 \ell} \cdots \alpha_{n-1}^{\lambda_{n-1} \ell/p} \right) \gamma_{n-1}^{m_{n-1}} = 0 (0 \leqslant m_{n-1} \leqslant L) ,$$

and since the Vandermonde determinant of order $L + 1$ with $\gamma_{n-1}^{m_{n-1}}$ in the $(\lambda_{n-1} + 1)$th row and $(m_{n-1} + 1)$th column is not 0, the new sums in parenthesis above would all vanish. By repeated application of the argument we would obtain

$$\sum_{\lambda_{-1}=0}^{L_{-1}} \sum_{\lambda_o=0}^{L_o} p(\lambda_{-1}, \ldots, \lambda_n) q(\lambda_{-1}, \lambda_o, \lambda_n, \ell/p)$$

$$= 0 (0 \leqslant \lambda_1 \leqslant L, \ldots, 0 \leqslant \lambda_n \leqslant L) .$$

Now if this equation were to hold for all integers ℓ, not divisible by p, with $1 \leqslant \ell \leqslant hk$, and all m_o with $0 \leqslant m_o \leqslant L$, then the function

$$\psi(x) = e^{\lambda_n\beta_o x}\sum_{\lambda_{-1}=0}^{L_{-1}}\sum_{\lambda_o=0}^{L_o} p(\lambda_{-1},\ldots,\lambda_n)(\Delta(\lambda_{-1}+x;h))^{\lambda_o+1}$$

would have at least

$$(hk - [hk/p])(L + 1) > h(L_o + 1)$$

zeros counted with multiplicities. But then also the polynomial $e^{-\lambda_n\beta_o x}\psi(x)$ would have this number of zeros, and since its degree is at most $h(L_o + 1)$, it follows that $\psi(x) = 0$ identically. From Lemma 2 we see that the polynomials

$$(\Delta(\lambda_{-1} + x;h))^{\lambda_o+1} \quad (0 \leqslant \lambda_{-1} \leqslant L_{-1},\ 0 \leqslant \lambda_o \leqslant L)$$

are linearly independent over the rationals, and thus $p(\lambda_{-1},\ldots,\lambda_n)$ would be 0 for all $\lambda_{-1},\ldots,\lambda_n$. The contradiction proves the assertion.

Hence we conclude that (14) holds for some integer ℓ, not divisible by p, with $1 \leqslant \ell \leqslant hk$, and some numbers $j_1,\ldots,j_{n-1}$ and α' as indicated above. This gives

$$\alpha_n = \alpha_1^{j_1}\ldots\alpha_{n-1}^{j_{n-1}}\alpha_n'^{p},$$

where $\alpha_n' = \alpha'^{1/\ell}$ for some ℓth root. In particular we see that $\alpha_n'^{p}$ is an element of K whence, since $(\ell,p) = 1$, it follows that α_n' must itself be an element of K. We assume now, as we may without loss of generality, that $\alpha_1 = -1$, and we assume also that $\log\alpha_1$ has its principal value $i\pi$; then

$$\log \alpha_n = (j_1+j)\log \alpha_1+\ldots+j_{n-1}\log \alpha_{n-1} + p \log \alpha'_n ,$$

where $\log \alpha'_n$ has its principal value and j is a rational integer with absolute value at most $c_{34}p$. On substituting for $\log \alpha_n$ in (1) we obtain

$$0 < |\beta'_o + \beta'_1 \log \alpha_1+\ldots+\beta'_{n-1}\log \alpha_{n-1} - \log \alpha'_n|$$
$$< (B \log A)^{-C \log A} ,$$

where

$$\beta'_o = \beta_o/p,\ \beta'_1 = (\beta_1-j_1-j)/p,\ \beta'_r = (\beta_r-j_r)/p(1 < r < n) .$$

Clearly $\beta'_o,\ldots,\beta'_{n-1}$ lie in K, which has degree $D \leqslant d^{2n}$. Further, the heights of $p\beta'_o,\ldots,p\beta'_{n-1}$ are plainly at most $c_{35}p^D B$ and since $p \leqslant 2k^{1/2}$, it follows that the heights of $\beta'_o,\ldots,\beta'_{n-1}$ are at most $c_{36}k^D B$. Furthermore, we observe that each conjugate of

$$\alpha'^p_n = \alpha_n\alpha_1^{-j_1}\ldots\alpha_{n-1}^{-j_{n-1}}$$

has absolute value at most $c^p_{37} A$, and the same estimate holds for some integer a such that $a\alpha'^p_n$ is an algebraic integer. Thus from Lemma 4 we deduce that the height of α'_n is at most $c_{38}A^{2D/p}$. Since $p > k^{1/2}$ we have $2D/p < 1/2$ and this confirms the assertions made at the beginning of §3.

The proof of the theorem is now completed by induction. First we note that there is no loss of generality in assuming, as in §3, that $\beta_n = -1$; for, if $\beta_n \neq 0$, then the height of β_r/β_n is at most B', where $\log B'/\log B$ is bounded above by a number depending only on d. Secondly we can suppose that d is taken, at the outset, as the degree of K so that the values of c_1, c_2 remain unaltered in the inductive discussion. Thirdly, there is plainly no loss of generality in assuming that $B \geqslant c_1^4$; to begin with we assume that also $A \geqslant c_2^4$, whence (1) remains valid with $c_1 B$ and $c_2 A^{1/2}$ substituted for B and A respectively. Then we can repeat the above argument and obtain for each $s = 1,2,\ldots$ a set of algebraic numbers $\beta_0^{(s)},\ldots,\beta_n^{(s)}$ with absolute values at most $c_1^s B$ and an element $\alpha_n^{(s)}$ of K with height at most

$$c_2^{1+1/2+\ldots+(1/2)^{s-1}} A^{(1/2)^s}$$

such that (1) holds with $\beta_0^{(s)},\ldots,\beta_{n-1}^{(s)}$ and $\alpha_n^{(s)}$ in place of $\beta_0,\ldots,\beta_{n-1}$ and α_n respectively. The algorithm terminates for some $s \leqslant 2 \log\log A$ when the height of $\alpha_n^{(s)}$ is at most c_2^4. Now there are only finitely many algebraic numbers with bounded degree and height, and thus the process can be continued to yield a number c, independent of A, such that $\alpha_n^{(j)} = \alpha_n^{(j')}$ for distinct $j,j' \leqslant c+2\log\log A$. This gives

$$\alpha_n^{(j)q} = \alpha_1^{p_1}\ldots\alpha_{n-1}^{p_{n-1}},$$

where $p_1,\ldots,p_{n-1},q$ denote rational integers with absolute values at most $p^{c+2\log\log A}$; on substituting for $\log \alpha_n^{(j)}$ in the inequality derived from (1) after j steps we obtain a linear form in which $\beta_n = 0$ and $\beta_0,\ldots,\beta_{n-1}$ are algebraic numbers with heights at most

$(\log A)^{c_{38}}B$. After repeating the argument n times we derive an inequality involving only β_0, and recalling that, by hypothesis, $\Lambda \neq 0$, this is plainly untenable. The contradiction proves the theorem.

[1] A. Baker, Linear forms in the logarithms of algebraic numbers, I, II, III, IV, Mathematika, 13 (1966), 204-216, 14 (1967), 102-107, 220-228, 15 (1968), 204-216.

[2] ________, Contributions to the theory of diophantine equations. I: On the representation of integers by binary forms, Philos. Trans. Roy. Soc. London A 263 (1968), 173-291.

[3] ________, A sharpening of the bounds for linear forms in logarithms, Acta Arith. 21 (1972), 117-129.

[4] ________, A sharpening of the bounds for linear forms in logarithms II, Acta Arith. (to appear).

[5] ________ and H.M. Stark, On a fundamental inequality in number theory, Annals of Math. 94 (1971), 190-199.

[6] N.I. Feldman, An improvement of the estimate of a linear form in the logarithms of algebraic numbers, Mat. Sb. 77 (119) (1968), 423-436.

[7] ________, An inequality for a linear form in the logarithms of algebraic numbers; Refinement of two effective inequalities of A. Baker, Mat. Zametki 6 (1969), 681-689, 767-769.

[8] ________, An effective sharpening of the exponent in Liouville's Theorem, Izv. Akad. Nauk SSSR. ser. mat. 35 (1971), 973-990.

[9] V.G. Sprindžuk, A new application of p-adic analysis to the representation of numbers by binary forms, Izv. Akad. Nauk SSSR, 34 (1970), 1038-1063.

[10] ________, On rational approximations to algebraic numbers, Izv. Akad. Nauk SSSR, 35 (1971), 991-1007.

Trinity College, Cambridge, England

THE MARKOV SPECTRUM

R. T. Bumby

0. Introduction, Notation and Terminology.

This paper is intended to put together the results from some recent papers [4,5,6,7] with some improvements. In addition we will explore the rumor that the sum with itself of the set of numbers with partial quotients bounded by 3 contains an interval. We would like to call attention to the fact that these methods are quite constructive and show that further improvements of the results are obtainable if they are desired. This also shows that the division between two types of behavior in the Markov spectrum is given by a recursive real number, although it says nothing about the arithmetic properties of that number. One suspects that the divisions will occur at quadratic irrationals and that these numbers could be written explicitly.

Let $B = \langle b_n : n \in Z, n \geq 0\rangle$ be a _sequence_ of positive integers, then $[B] = [b_0, b_1, \ldots]$ will refer to the number represented by the _continued fraction_ whose _partial quotients_ are the b_n. Thus $[c,B] = c + 1 \div [B]$. If $A = \langle a_n : n \in Z\rangle$

is a <u>two-ended sequence</u> of positive integers we define $A_n^+ = \langle a_{n+i} : i \in Z, i \geq 0 \rangle$ and $A_n^- = \langle a_{n-i} : i \in Z, i \geq 0 \rangle$. Then define

$$M_n(A) = [A_n^+] + [A_n^-] - a_n$$

$$M(A) = \sup\{M_n(A) : n \in Z\}.$$

The <u>Markov spectrum</u> is the set of values of $M(A)$ as A runs through all two-ended sequences of positive integers. This quantity $M(A)$ can be used to express the minimum of an inhomogeneous quadratic form or

$$L(\alpha) = \lim\inf\{|q(q\alpha-p)| : p,q \in Z, q \neq 0\} .$$

Now $\{L(\alpha) : \alpha \in R\} \subsetneq \{1 \div M(A)\}$, but there are sequences A which do not correspond to any α. The fact that L is defined by the lim inf instead of the inf causes this phenomenon. For all matters related to the question of proper containment of these sets we refer to Berstein [1] and his references. Related questions may also be found in the paper of Diviš [3]. Having indicated that the problem belongs to Diophantine Approximation, we may now ignore the connection and concentrate on the continued fraction method.

Our technique is based on the construction as a <u>Cantor set</u> of the set of numbers whose continued fractions have certain properties. A <u>Cantor set</u> X is given by constructing a sequence of covers $C_0(X)$, $C_1(X)$,... such that

1) each C_i is a union of $i + 1$ closed intervals
2) C_{i+1} is obtained from C_i by removing an open interval G_i from the interior of some interval I_i turning it into two intervals I_i' and I_i''
3) $X = \bigcap C_i$

Many different constructions will yield the same X. The complement of X in C_0 is the union of the disjoint open intervals G_i, so X determines $\{G_i\}$ but not the order in which we take them. Since there can be only finitely many G_i of length $>\varepsilon$, we could order them by decreasing length. This will be the <u>standard construction</u> of X.

We will be working with sets which are defined by specifying the first few partial quotients. Then one expands $[b_0, \ldots, b_n, x]$ into the form $(p_n x + p_{n-1}) \div (q_n x + q_{n-1})$. Replacing x by $(b_{n+1} + 1 \div x)$ and simplifying gives the same form with $p_{n+1} = p_n b_{n+1} + p_{n-1}$, $q_{n+1} = q_n b_{n+1} + q_{n-1}$. Starting from [x] we may take $p_{-1} = 1$, $p_{-2} = 0$, $q_{-1} = 0$, $q_{-2} = 1$; $p_{n+1} q_n - p_n q_{n+1} = p_{n-1} q_n - p_n q_{n-1}$, so $p_{n+1} q_n - p_n q_{n+1} = (-1)^n$. Thus

$$\frac{p_n x + p_{n-1}}{q_n x + q_{n-1}} - \frac{p_n y + p_{n-1}}{q_n y + q_{n-1}} = \frac{(-1)^{n-1}(x-y)}{(q_n x + q_{n-1})(q_n y + q_{n-1})}$$

Also $q_n \div q_{n-1} = [b_n, \ldots, b_1]$. In addition there will be some restriction on the partial quotients that we allow after those we specify. X will denote the set of numbers between 0 and 1 that are allowed, and $X[b_1,\ldots,b_n]$ will denote those which have continued fractions beginning with $<0,b_1,\ldots,b_n>$.

The early history of the Markov spectrum can be found in Koksma's book [7]. Markov proved that the portion of the spectrum below 3 consists of a sequence of values with 3 as their only limit point. Cohn [2] has some contributions to this part of the spectrum and a discussion of Markov's method. However, we will be looking at the gaps in the spectrum above 3.

1. All Numbers Greater Than 5.1 Belong to the Markov Spectrum.

A classical result of Hall states:

PROPOSITION. If X and Y are two Cantor sets whose standard constructions satisfy

(H) at each stage of the construction G_i is shorter than both I_i' and I_i''

and such that $C_0(X)$ is longer than $G_0(Y)$ and $C_0(Y)$ is longer than $G_0(X)$ then $X \dot{+} Y. = C_0(x) + C_0(Y)$.

Proof. One forms a sequence of covers of the cartesian product $X \times Y$ of the form $C_i(X)\times C_j(Y)$ where at each stage we remove the larger of $G_i(X)$, $G_j(Y)$. This converges to $X \times Y$. We assume inductively that $C_i(X) + C_j(Y) = C_0(X) + C_0(Y)$ and that each interval of $C_i(X)$ is longer than $G_j(Y)$ and each interval of $C_j(Y)$ is longer than $G_i(X)$. If we are to remove $G_i(X)$ at this stage we must show

(i) $(I_i'(X) \cup I_i''(X)) + J = I_i(X)+J$ where J is an interval of $C_j(Y)$

(ii) each interval of $C_j(Y)$ is longer than $G_{i+1}(X)$

(iii) each interval of $C_{i+1}(X)$ is longer than $G_j(Y)$

From (H), I_i', I_i'' are each longer than $G_i(X)$, which by choice is longer than $G_j(Y)$. The rest of (iii) is already known; (ii) depends only on $G_{i+1}(X)$ being shorter than $G_i(X)$; (i) is a consequence of J being longer than $G_i(X)$.

If (H) holds for any construction, then it holds for the standard construction so intrinsic versions of this theorem can be formulated.

Freiman and Yudin [4] apply this to a set F of numbers whose partial quotients are all ≤ 4 and for which a partial quotient of 4 only arises after certain strings of partial quotients. Hall [5] used the same set, but gave a slightly sharper application to the Markov spectrum. We modify the set of Freiman and Yudin slightly to obtain a still sharper result. For our F, after the sequence $b_0, b_1, \ldots, b_n$ we allow 4 only if $k_n = [b_n, \ldots, b_1] > [\overline{3}]$ (<u>where the overscore</u>

<u>indicates that that portion is to be repeated indefinitely)</u>. For these purposes $k_0 = [\ ] = \infty$, so that $[b,\ ] = b+1\div[\ ] = b+1\div\infty = b$.

If $k_n > [\overline{3}]$, then we allow b_{n+1} to be 1,2,3, or 4. We form the convex hull of possible values of $[b_{n+1}, b_{n+2}, \ldots]$ which we call a 1-4 interval. We split this into two parts depending on whether $b_{n+1} = 1$ or not. The latter we call a 2-4 interval which we split into two parts depending on whether $b_{n+1} = 2$ or not. We then get a 3-4 interval which we split according to whether $b_n = 3$ or 4. After $b_{n+1} = 1,2,$ or 3, $k_{n+1} = b_{n+1}+1\div k_n < [\overline{3}]$, so b_{n+2} cannot be 4 and we get a 1-3 interval. After $b_{n+1} = 4$ we get another 1-4 interval. The 1-3 interval splits according to $b_{n+1} = 1$ or not. The latter is a 2-3 interval which splits according to $b_{n+1} = 2$ or 3. Now if $b_{n+1} = 3$, $k_{n+1} > [\overline{3}]$ so we get a 1-4 interval. After 1 or 2 we get a 1-3 interval. We must verify that this construction satisfies (H). The ratio of the lengths of the interval joining $[b_0, \ldots, b_n, \alpha]$ to $[b_0, \ldots, b_n, \beta]$ and the interval from $[b_0, \ldots, b_n, \beta]$ to $[b_0, \ldots, b_n, \gamma]$ is

$$\frac{\alpha-\beta}{\beta-\gamma}\,\frac{\gamma+r}{\alpha+r} \quad \text{where} \quad r = q_{n-1}\div q_n = [0, b_n, \ldots, b_1]$$

We may assume that $0 \leqslant r \leqslant 1$, and we note that this ratio is a monotonic function of r; increasing if $\alpha > \gamma$, decreasing if $\alpha < \gamma$.

The data on endpoints and division points is given in Table 1.

type	end	division		end
1-4	$[\overline{1,3}]$	$[1,\overline{1,3}]$	$[2,\overline{3,1}]$	$[4,\overline{1,3}]$
2-4	$[2,\overline{3,1}]$	$[2,\overline{1,3}]$	$[3,\overline{3,1}]$	$[4,\overline{1,3}]$
3-4	$[3,\overline{3,1}]$	$[3,\overline{1,3}]$	$[4,4,\overline{1,3}]$	$[4,\overline{1,3}]$
1-3	$[\overline{1,3}]$	$[1,\overline{1,3}]$	$[2,\overline{3,1}]$	$[\overline{3,1}]$
2-3	$[2,\overline{3,1}]$	$[2,\overline{1,3}]$	$[3,4,\overline{1,3}]$	$[\overline{3,1}]$

Table 1

Some constants we need are $[0,\overline{1,3}] = \frac{\sqrt{21}-3}{2}$, $[0,\overline{3,1}] = \frac{\sqrt{21}-3}{6}$, $[0,4,\overline{1,3}] = \frac{5-\sqrt{21}}{2}$. The intervals at the left of our table are already longer than the middle intervals and the formula calls for multiplying this ratio by a quantity greater than 1, so there is no difficulty here. For the right end÷middle we have an increasing function of r. Among the 1-4, 2-4, and 1-3, the smallest value will be obtained by putting $r = 0$ in the ratio for the 1-3 interval which gives

$\frac{\sqrt{21}}{6-\sqrt{21}} \times \frac{\sqrt{21}-1}{\sqrt{21}+3} = \frac{\sqrt{21}}{3} > 1$. Similarly, between the 3-4 and 2-3 intervals, the latter gives the smaller value. Evaluating at $r = 0$ gives

$\frac{\sqrt{21}-4}{5-\sqrt{21}} \times \frac{\sqrt{21}+1}{\sqrt{21}+3} = \frac{17-3\sqrt{21}}{2\sqrt{21}-6} > 1$ (since $17-3\sqrt{21} > 2\sqrt{21}-6 \iff \sqrt{21} < 4.6 \iff 21 < 21.16$). This construction then satisfies (H). The elements <1

in this set form a Cantor set whose sum with itself is the interval from $5-\sqrt{21}$ to $\sqrt{21}-3$ which has length greater than 1. Now take all values of $a_0 \geq 4$ and choose $[A_0^+] - a_0$, $[A_0^-] - a_0 \in F$. Then for different choices of a_0, $M_0(A)$ will run through all values $\geq 9-\sqrt{21}$. But, it is still possible that $M(A) > M_0(A)$. In other words, we are limited by the sup of the possible values of $M_n(A)$, $n \neq 0$. If $a_n \neq 4$, $M_n(A) \leq \sqrt{21}$, but if for some $n > 0$, $a_n = 4$, then $k_n > [\bar{3}]$ and hence $[A_n^-] \leq [4,\bar{3}]$. Meanwhile $[A_n^+] \leq [4,\overline{1,3}]$. Thus, we have

<u>THEOREM</u>. The Markov spectrum contains all values $\geq [4,\overline{1,3}] + [0,\bar{3}] = \frac{5+\sqrt{21}}{2} + \frac{\sqrt{13}-3}{2} = 5.094\ldots$

It appears that this may be the best that can be done via property (H). If we allowed 4 to be both preceded and followed by 1 we would have $M(A) \geq M(\overline{4,1,1}) = \sqrt{26}$ which is larger. If we want to rule out 4,1 after 1 then the division of the 3-4 interval will not satisfy (H). We are left with forbidding 4 after 1. This leads to a smallest possible value of $[\overline{1,3}]$. When a 4 occurs we must allow it to be followed by 1,3 for we cannot have an interval satisfying (H) whose first terms are only 1 or 2. The interval $[3,4,\overline{1,3}] - [\overline{3,1}]$ cannot be appreciably shortened since its largest value must begin with 3,1,3,... This forces periodicity. This brief summary masks a fair amount of computation.

2. The Hausdorff Dimension of the Portion of the Spectrum Less Than α.

Both the papers of Kinney and Pitcher [7] and of Hall [5] contain proofs of the fact that the portion of the spectrum below $\sqrt{10}$ has measure 0. Kinney and Pitcher point out that the proof actually shows that the Hausdorff dimension is less than 1. Both of these proofs are based on calculations. Kinney and Pitcher fail to use the best bounds on the intervals they are considering but they observe that one can assume that n is large and they compute the bounds on $q_n \div q_{n-1}$ on this basis. On the other hand, Hall uses less precise bounds on $q_n \div q_{n-1}$ but is more precise about the intervals in the Cantor set construction. Although the result we are about to prove could be obtained by similar calculation, we prefer to use a small amount of the theory of Hausdorff measure for which we refer to Rogers' book [9] and the papers cited therein. The crucial result is that the cartesian product of two sets will have dimension equal to the sum of the dimensions of the factors under appropriate conditions which happen to be satisfied here. The portion of the Markov spectrum below a certain value is contained in a finite union of translates of a set of the form $X + X$ where X is the set of all numbers between 0 and 1 whose partial quotients satisfy certain restrictions. This set is a projection of the cartesian product $X \times X$, so if $X \times X$

has dimension less than 1, so does $X + X$. Invoking the theorem, it becomes necessary to decide how the dimension of X compares with $\frac{1}{2}$. If less, then $X + X$ must have measure 0; if greater, it is necessary to study the projection $X \times X \to X + X$ to obtain further information. Although these questions are of considerable interest (e.g. if $\dim X > \frac{1}{2}$, will $X + X$ contain an interval?) we shall limit ourselves to calculating the dimension of $X \times X$.

If we require that partial quotients be bounded by 2 then we are dealing with numbers between $[\overline{1,2}] = \frac{1+\sqrt{3}}{2}$ and $[\overline{2,1}] = 1+\sqrt{3}$. The intervals in our Cantor set construction will be linear fractional transforms of this interval, and they are always divided at the transforms of $[1,\overline{1,2}]$ and $[2,\overline{2,1}]$. The ratios of the intervals to the whole will be greater than

$$\frac{[1,\overline{1,2}]-[\overline{1,2}]}{[\overline{2,1}]-[\overline{1,2}]} \quad \frac{[\overline{2,1}]+[0,\overline{1,2}]+\varepsilon}{[1,\overline{1,2}]+[0,\overline{1,2}]+\varepsilon}$$

and

$$\frac{[\overline{2,1}]-[2,\overline{2,1}]}{[\overline{2,1}]-[\overline{1,2}]} \quad \frac{[\overline{1\ 2}]+[0,\overline{2,1}]-\varepsilon}{[2,\overline{2,1}]+[0,\overline{2,1}]-\varepsilon}$$

when we begin with a large enough value of n. Actually, r must be the same for both pieces, so when one is small, the other cannot be. If the sum of the square roots of these quantities is greater than one, then refining a cover C_i

must increase the value of the sum of the square roots of the lengths of the intervals. Thus the $\frac{1}{2}$-dimensional measure is infinite. Computation shows that it is. In fact, just using these lower bounds, and similar upper bounds we find that the dimension is between .51975 and .54322, considerably sharpening the bounds of 1/3 and 2/3 given in Rogers' book. (see Appendix).

To show that the cartesian product has dimension at least the sum of the dimensions we give a lower bound on the $\alpha+\beta$ dimensional measure by taking the product of the α power of the length of the first projection and the β power of the length of the second projection of a set in a cover. The sum of this expression over the cover can be in turn bounded below by the product of suitable expressions on the factors.

The value $\sqrt{10}$ occurring in previous proofs arises from looking at sequences whose elements are bounded by 2 and which do not contain $<1,2,1>$. The least value of $M(A)$ for which A contains $<1,2,1>$ is $M(\overline{2,1,1}) = \sqrt{10}$. A similar result which we need to extend these results is given below.

<u>PROPOSITION</u>. If we allow B to contain the sequence $<1,2,1,2>$ then $M(B) \geq M(\overline{1,2,1,2,2,1,2,1}) = \frac{\sqrt{689}}{8}$.

<u>Proof</u>. Suppose that $b_{-1} = 1$, $b_o = 2$, $b_1 = 1$, $b_2 = 2$ and $M(B) \leq \sqrt{689} \div 8$. Then setting $x=[B_0^-]$,

$y = [B_0^+]$ we have $M_0(B) = x+y-2$. Thus either $x \leq [\overline{2,1,1,2,1,2,2,1}] = \alpha$ or $y \leq [\overline{2,1,2,2,1,2,1,1}] = \beta$. In the former case, x begins with $[2,1,1,2,1,2,\ldots]$; and we set $x = [2,1,1,u]$. Now $M_0 = y+f(u)$ and $M_{-3} = u+f(y)$ where $f(t) = \frac{t+1}{2t+1}$; $M_0 - M_{-3} = (y-f(y))-(u-f(u))$; $t-f(t)$ is an increasing function; hence $M_0 \geq M_{-3} \iff y \geq u$. By symmetry, there is no loss of generality in assuming $M_0 \geq M_{-3}$ in this context. Then, since $f(t)$ is a decreasing function, $M_0 \geq y+f(y)$. Hence $u \leq y \leq \beta$. Meanwhile x is a decreasing function of u so $x \geq \alpha$, thus $x = \alpha$. Starting from $y \leq \beta$, we get that y begins with $[2,1,2,2,1,2,1\ldots]$. Now we compare M_0 with M_5 with the result that $M_0 \geq M_5 \implies y \geq \beta$ and $x \leq \alpha$.

The portion of the Markov spectrum below $\frac{\sqrt{689}}{8}$ is determined by sequences in which $\langle 1,2,1,2 \rangle$ never occurs. The intervals in the Cantor set construction of this set are of four types described in Table 2.

Table 2

after	endpoints		bounds on 1/r	
...1,1	$[\overline{1,2,1,1}]$	$[\overline{2,1,1,1}]$	$[\overline{1,1,1,2}]$	$[\overline{1,1,2,1}]$
...2,2	$[\overline{1,2,2,2}]$	$[\overline{2,1,2,2}]$	$[\overline{2,2,1,2}]$	$[\overline{2,2,2,1}]$
...2,2,1	$[\overline{1,2,1,1}]$	$[\overline{2,2,2,1}]$	$[\overline{1,2,2,2}]$	$[1,\overline{2,2,1,2}]$
...1,1,2	$[\overline{1,1,1,2}]$	$[\overline{2,1,2,2}]$	$[\overline{2,1,1,2,1}]$	$[\overline{2,1,1,1}]$

Note that any occurrence of $<1,2,1>$ must be $<1,2,1,1>$, and $<2,1,2>$ must be $<2,1,2,2>$; so all types are covered by table 2. To show that the dimension of this set is $\leqslant \frac{1}{2}$, we show that every cover can be refined to one in which the sum of the square roots of the lengths of intervals is smaller. It suffices to work with the covers given by a Cantor set construction. These contain intervals which are images of those in table 2. The subdivision always occurs at the images of $[\overline{1,1,2,1}]$ and $[\overline{2,2,1,2}]$. What we actually show is that some finite number of subdivisions of an interval gives a smaller value to the sum of the square roots of the lengths. This actually shows measure 0 in some dimension α less than $\frac{1}{2}$. It also shows that we can obtain a smaller contribution to the α-dimensional measure by covers in which all intervals have length between ε and $c\varepsilon$ for some constant $c < 1$. The number $n(\varepsilon)$ of intervals in the cover satisfies $n(\varepsilon)\cdot\varepsilon^{\alpha} < c_2$ for constant c_2. The cartesian product of this with itself consists of $n(\varepsilon)^2$ rectangles of diameter less than $\varepsilon\sqrt{2}$. Its one dimensional measure is then bounded by $c_3 n(\varepsilon)^2\varepsilon = O(\varepsilon^{1-2\alpha})$. In fact the cartesian product will then have dimension at most 2α. The indicated calculation has been performed (see Appendix).

3. Gaps in the Spectrum Above $\frac{\sqrt{689}}{8}$.

The portion of the spectrum below $\sqrt{689} \div 8$ has been shown to have measure 0, so it can contain no interval. Above that value, there may be some short intervals, but there are infinitely many gaps. Kinney and Pitcher [7] give a general recipe for constructing them, but except near $\sqrt{12}$ they give no indication that their "gaps" are separated by points of the spectrum. Hightower [6] gives a slightly different version of this construction and proves that there are infinitely many distinct gaps approaching $1 + \sqrt{5}$ from either side. Again we combine results, and extend them beyond $\sqrt{12}$.

In this section we require all partial quotients to be bounded by 2. If we want to construct the sum of $X[b] = X[b_1,\ldots,b_n]$ with itself, we begin with the cartesian product and project by the map $(x,y) \to x+y$. Note that we have the same set on each factor. Perform the Cantor set construction on both factors. The square $C_0 \times C_0$ which covered the product is replaced by two squares and two rectangles at the first step. In fact, in this Cantor set construction, G_i is always longer than both I_i' and I_i'' (although shorter than their sum). Thus $X[b]+X[b]$ is covered by the three disjoint intervals $I_i'+I_1'$, $I_1'+I_1''$, $I_1''+I_1''$. The middle one we can't do anything about, but the other two are covers of the sum of some $X[b]$ with itself, so we can repeat the construction for them. This produces infinitely many gaps in $X + X$. But the Markov spectrum is somewhat smaller, so some of our intervals may not contain any points of the spectrum. This construction, then, should be combined with a procedure which gives us points of the spectrum in some of the intervals of our cover. The procedure will be applied to that portion of the spectrum where $\langle 1,2,1\rangle$ is allowed, i.e. the values greater than $M(\overline{1,2,1}) = \sqrt{10}$. Set $a_0 = 2$, $a_1 = 1$ and define $A = \{a_n : n \in z\}$ in such a way that $a_n = a_{-n}$, all $a_n = 1$ or 2, and $M(A) > M_0(A)$. Since we have shown that $X(d,1) + X(d,1)$, $X(d,1) + X(d,2)$, and $X(d,2) + X(d,2)$ are contained in disjoint intervals for any sequence d, it is easy to test whether $M(A) = M_0(A)$. We characterize sequences by the <u>length</u>, which is the greatest n such that $a_k \equiv k \pmod 2$ for all $0 \leq k \leq n$. Then $M(A) = M_0(A)$ if no more than $2n$ consecutive terms agree with $\langle\overline{1,2}\rangle$ except for $\langle a_{-n}, \ldots, a_n\rangle$. Suppose also then that A has this property. Hightower's sequence ($a_0 = 2$, $a_n = 1$ for $n \neq 0$) has this property, for example. Then we can find neighboring sequences of the same type by the following construction: (1) if $a_{n+1} \neq a_n$, set all $a'_k = a_n$ for $|k| \geq n$ and $a'_k = a_k$ for $|k| \leq n$; (2) if $a_{n-1} = a_n = a_{n+1}$, set $a'_k = 3 - a_n$ for $|k| > n$ and $a'_k = a_k$ for $|k| \leq n$. It is clear that $A' = \{a'_k : k \in Z\}$ has the required property for each value of n for which this construction is allowed. There are infinitely many such n. These A' are separated from each other by the gaps we just constructed.

Now we look beyond $\sqrt{12} = 3.46410\ldots$. It has been known for a long time [8] that the next largest values in the spectrum are $M(\overline{3}) = \sqrt{13} = 3.60555$ and $M(\overline{1,2},3,3,\overline{2,1}) = \dfrac{65+9\sqrt{3}}{22} = 3.66311\ldots$. We will produce infinitely many gaps between here and $M(\overline{3,2,2,3}) = \dfrac{\sqrt{3965}}{17} = 3.70401\ldots$. This quantity was chosen on the basis of the computation of the smallest value of $M(A)$ for A which contain certain finite sequences S (and not some others known to give larger values). The results of these computations are given in Table 3. For convenience, we drop all punctuation between the elements of the sequences A_0 and S. We give the first 5 places of the decimal expansion. Here A_0 is the sequence for which $M(A)$ takes the smallest value.

Table 3

S	A_0	$M(A_0)$
<13>	$(\overline{3113})$	3.84654
<232>	$(\overline{12}3\overline{21})$	3.73205
<323>	$(\overline{323})$	3.72525
<322>	$(\overline{3223})$	3.70401
<3211>	$(\overline{21}1123321\overline{12})$	3.67554
<3332>	$(\overline{12}3333\overline{21})$	3.66875

Table 3 (continued)

S	A_0	$M(A_0)$
<32123>	$(\overline{32123})$	3.66741
<32122>	$(\overline{12}22123321222\overline{21})$	3.66609
<321211>	$(\overline{21}12123321211\overline{12})$	3.66453

As a companion to Table 3 we have the largest values when all S above the stated one in the table are forbidden from A. The largest value will be $M(A_1)$. These values are given in Table 4.

Table 4

S	A_1	$M(A_1)$
<13>	$(\overline{32})$	3.87298
<232>	$(\overline{3332})$	3.73811
<323>	$(\overline{12}23332\overline{21})$	3.72640
<322>	$(\overline{21}1233321\overline{12})$	3.69147
<3211>	$(\overline{321233})$	3.67408
<3332>	$(\overline{12}33212332\overline{21})$	3.66783
<32123>	$(\overline{12}33212\overline{21})$	3.66705
<32122>	$(\overline{12}332121\overline{12})$	3.66467
<321211>	$(\overline{12}3321212332\overline{21})$	3.66344

The tables show gaps on all but the first three lines of the tables. The last three lines are typical of the gaps we can produce. For example, we form the smallest element containing 32...123 but not 32...111 and compare it to the largest element not containing 32...123. These have the forms $M(\overline{32...123})$ and $[0,3,\overline{2,1}]+[3,2...1,2,\overline{2,1}]$. Let $f(t) = [3,2,...,1,2,t]$. Then we have a gap if $f(x)+1 \div f(x) > f(y)+1 \div f(z)$ where $x = [\overline{3,3,2,...,1,2}]$, $y = [\overline{2,1}]$, $z = [\overline{1,2}]$. This reduces to showing that

$$\frac{f(x)-f(y)}{f(x)-f(z)} > \frac{1}{f(x)f(z)} .$$

The left side reduces to

$$\frac{x-y}{x-z} \; \frac{z+r}{y+r}$$

where $r = [0,2,1,..,2]$. This is an increasing function of r and $r \geqslant [0,\overline{2,1}]$. It is also an increasing function of x and $x \geqslant [3,3,\overline{2,1}]$. The left side is then greater than .113. Meanwhile the right side is smaller than $1 \div [3,\overline{2,1}]^2 = .088$. The other gaps have the same form with x replaced by $[2,\overline{2,1}]$ and y by $[1,\overline{1,2}]$ or with x replaced by $[1,\overline{1,2}]$ and y replaced by $[1,2,3,3,\overline{2,1}]$. The left side in these two cases are .523 and .783.

Between 3.704 and 3.725, we have ruled out <3,2,3> and we must have <3,2,2> in A. We have also ruled out any occurrence of <1,3> or <2,3,2>. In addition all occurrences of <3,2,1> will not be at the maximum value of $M_n(A)$ for our calculation of A_1 for <3,2,2> depended only on the fact that <3,2,2> did not occur at the maximum $M_n(A)$ and <3,1> did not occur at all. The least M when <3,3,3,2,2>

is allowed is $M(\overline{3,3,3,2,2}) = 3.71445$; the greatest when it is excluded is $M(\overline{1,2},3,3,2,\overline{2,1}) = 3.71973$; so these intervals overlap. The least M when $\langle 3,2,2,2\rangle$ is allowed is $M(\overline{1,2},2,2,3,3,2,2,\overline{2,1}) = 3.70578$; the least when $\langle 1,2,3,3,2,2\rangle$ is allowed is $M(\overline{2,1},1,2,3,3,2,2,3,3,2,1,\overline{1,2}) = 3.70598$. When these are forbidden, the only value coming from $\langle 3,2,2\rangle$ is $M(\overline{3,2,2,3}) = 3.70401$. This value is then isolated. Computations have not led us to any gap in the spectrum beyond this. We have, in fact, found that several intervals in a cover of the spectrum overlap. The gaps we have found so far end at $(27579+265\sqrt{3})\div 7566 = 3.70578\ldots$.

4. The Sum Set of Numbers with Partial Quotients Bounded by 3.

Property (H) fails for the set of numbers with partial quotients bounded by 3. The interval of elements that begin with 3 is too short, but this seems to be the only failing. Elsewhere, property (H) holds quite strongly. But the proof of Hall's theorem demands that it hold everywhere; so one must analyze the process more carefully if one is to prove that almost property (H) implies that the sum set is almost the whole interval.

We are now working with partial quotients bounded by 3, except for application to the Markov spectrum where we will need the set $X[4]$ of numbers whose partial quotients are ≤ 3 except for the first one which is 4.

We want to study the gaps in $X[a] + X[b]$, so we form covers C, D and look for gaps in $C + D$. All gaps can be found in this way. Let α be a fixed value not in $C + D$. We may assume that C and D are actually steps in the Cantor set constructions of $X[a]$ and $X[b]$. Our procedure will essentially be an induction on the numbers of intervals in C.

If C contains only one interval, the gaps in $C + D$ are given by those gaps in D which are longer than C. Now assume that C has $n + 2$ intervals, and look at the intervals I'_n and I''_n created in the last step of the construction of C, and suppose that I'_n contains the smaller values. The line $x+y = \alpha$ contains no points of $C \times D$, so it must pass through gaps H' and H'' in D over I'_n and I''_n, respectively (see Figure 1). If $H' = H''$, then there is no problem: we can simply refrain from performing the last step of the construction since I_n+D already does not contain α. Otherwise, we start undoing the steps in the construction of D until we would have to fill either H' or H''. This does not change the appearance of Figure 1. Suppose now that H' is the last interval removed in D, then $H' > I'_n$ and $G_n > J'$. Thus $\frac{I'_n}{G_n} \cdot \frac{J'}{H'} < 1$. If we had property H, this would be impossible. As it is, it can hold only in special cases. (Our choices are just for definiteness in the figure.

Other choices give symmetrical results).

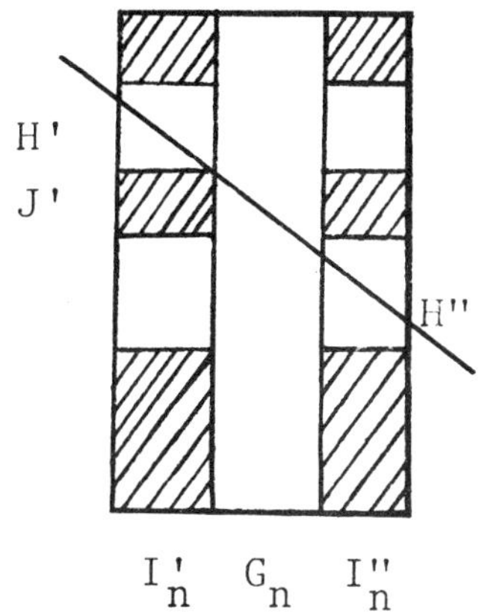

Figure 1

The ratios $I'_n \div G_n$ and $J' \div H'$ can be bounded independently. Their product can be less than 1 only if both I'_n and J' are the 3-intervals resulting from division of 2-3-intervals in the usual construction of these Cantor sets. In this case, it is also clear that there could be at most one interval between H' and H" (when H' is the last gap). Now we recombine I_n in exchange for allowing the cartesian products of two 2-3-intervals to cross the line $x+y = \alpha$ as it would be required to do in order to produce Figure 1. We assume that the modification in D has been made only over I_n, so we no longer have a cartesian product. At most a certain fraction (slightly more than half) of the 2-3-interval is allowed to cross $x+y = \alpha$. We then return to our induction with this new possibility. The first place where this changes the procedure is where we are attaching a 1-interval to one of these 2-3-intervals. Here we get no new crossing of $x+y = \alpha$, but now we must allow intervals of either 1,2, or 3 type to cross the line near one end. Continuing these constructions we find that we must allow crossing near either end. Now the trouble begins since we can run into the construction leading to Figure 1 in which all rectangles near the line are actually allowed to cross it slightly. This weakens the inequalities allowing more crossing, etc. These procedures will give us a catalog of the types of gaps in $X[a] + X[b]$. When we suspect that the catalog is complete, we set up an induction to prove that the types of gaps we have allowed will not lead to any new gaps. This procedure also leads to estimates on the size of the largest interval in $X[a] + X[b]$.

Ratios of images of intervals with a common point are given by expressions of the form

$$\frac{a-b}{a-c} \quad \frac{c+r}{b+r}$$

which is a cross ratio involving the four values $a,b,c,-r$. If we choose $[\overline{1,3}] - [\overline{3,1}]$ and $[2,\overline{3,1}]$ $[\overline{3,1}]$ as our prototype intervals, $0 \geq -r \geq -1$. Any change in the prototype by a projective transformation will lead to the same computational results if the range of $-r$ is subjected to the same change. We choose to normalize these intervals so that the first interval removed is in the same place. Subtracting appropriate quantities gives us

$$[0,\overline{3,1}]-[0,\overline{1,3}] \quad [\overline{1,3}]-[2,\overline{1,3}] \quad -1\geq -r\geq -2$$

and

$$[0,\overline{3,1}]-[0,\overline{1,3}] \quad [\overline{1,3}]-[1,\overline{1,3}] \quad -2\geq -r\geq -3 \ .$$

Intervals arising in our analysis need not always have the same orientation, so we add to these their inverses:

$$[0,2,\overline{1,3}]-[0,\overline{1,3}] \quad [\overline{1,3}]-[\overline{3,1}] \quad -\tfrac{1}{2}\geq -r\geq -1$$

$$[0,1,\overline{1,3}]-[0,\overline{1,3}] \quad [\overline{1,3}]-[\overline{3,1}] \quad -\tfrac{1}{3}\geq -r\geq -\tfrac{1}{2} \ .$$

A combination of the last G_i in the construction of our cover and its adjacent I_i' can be given by

$$[0,\overline{1,3}] \quad [\overline{1,3}] \quad [1,\overline{1,3}] \quad -2\geq -r\geq -3$$

$$[0,\overline{1,3}] \quad [\overline{1,3}] \quad [2,\overline{1,3}] \quad -1\geq -r\geq -2$$

$$[0,\overline{1,3}] \quad [\overline{1,3}] \quad [3,\overline{1,3}] \quad -\tfrac{1}{3}\geq -r\geq -1 \ .$$

The first of these intervals represents a 3 interval in a 2-3-interval. The second, a 2-3 interval in a 1-3-interval. Here $[1,\overline{1,3}]$ represents the end of the 2-interval. The third represents either a 2-interval in a 2-3-interval or a 1-interval in a 1-3-interval. $[1,\overline{1,3}]$ and $[2,\overline{1,3}]$ are the ends of the 1 and 2 subintervals when this interval is partitioned. To embed this in one of our prototypes we proceed as follows:

from case 1, add 1 to embed in case 2;

from case 2, add 1;

either now in case 3 or we have a 3-interval;

from case 3 or a 3-interval, apply $1+1\div x$;

3 interval goes to case 1;

case 3 goes to case 2 or 3 or a 3-interval.

The 3-interval ending at $[\overline{1,3}]$ has a gap which goes beyond $[0,\overline{1,3}]$. The values of $-r$ for a 3-interval are between $-\frac{1}{4}$ and $-\frac{1}{3}$. In our case we find that a rectangle can split off only if both sides belong to case 1. Add 2, invert, and add 1 to get this interval placed in a new prototype in the veritcal direction. Horizontally we move step by step. When we duplicate Figure 1, only that portion surrounding H' need concern us. Actually moving as in the vertical direction causes no difficulty provided we recall which questions we should be asking. Any time that we have reduced ourselves to handling a 3-interval, we invert, add 3, invert and add 1 before proceeding. This moves the whole picture closer to $[\overline{1,3}]$. Very soon, however, small differences at this point produce no qualitative differences. To show that $X + X$ contains all values between $[0,\overline{3,1}]+[0,2,\overline{1,3}]$ and $[0,\overline{1,3}]+[0,1,2,\overline{1,3}]$ would require more detailed analysis of this process, but approximations to this result could be established by direct calculation. Sample computations are in the appendix. A more rigorour treatment of these results, improvements, and applications to the Markov spectrum should be undertaken.

REFERENCES

[1] A.A. Berstein, Necessary and sufficient conditions for the occurrence of points in the Markov spectrum in the Lagrange spectrum, Doklady, Akad. Nauk. SSSR 191 (1970) 971-973. English translation: Soviet Mathematics - Doklady 11 (1970) 463-466.

[2] H. Cohn, Representation of Markoff's binary quadratic forms by geodesics on a perforated torus, Acta Arith. 18 (1971) 125-136.

[3] B. Diviš, An analog of the Lagrange numbers, J. Number Theory 4 (1972) 274-285.

[4] G.A. Freiman, & A.A. Yudin, On Markovian spectrum (Russian) Litovskiĭ Mat. Sbornik 6 (1966) 443-447.

[5] M. Hall, Jr., The Markoff spectrum, Acta Arith. 18 (1971) 387-399.

[6] C.J. Hightower, The minima of indefinite binary quadratic forms, J. Number Theory 2 (1970) 364-378.

[7] J.R. Kinney and T. Pitcher, On the lower range of Perron's modular function, Can. J. Math. 21 (1969) 808-816.

[8] J.F. Koksma, Diophantische Approximationen, Ergebinisse der Math...4, Springer, Berlin 1936 (especially Chapt. III, Sec. 2).

[9] C.A. Rogers, Hausdorff Measures, Cambridge University Press, Cambridge 1970 (especially Chapt. 3, Sec. 2).

Rutgers University, New Brunswick, New Jersey

```
002)• 16.33.34 11/09/72 BUMBY

   A P L \ 3 6 0

      )LOAD MARKOV
SAVED  17.11.02 08/04/72
      )FNS
ADD     BRIDGE  COMP    CR      CYCLE   DER     DIMENSION
INV     M       NEWT    ORD     SECT4   TABLE2  VAL
      ∇M[⎕]∇
    ∇ M;X;N1;D1
[1]   'ENTER CYCLE'
[2]   CYCLE X←⎕
[3]   'ENTER FIRST BRIDGE'
[4]   BRIDGE X←⎕
[5]   N1←NUM
[6]   D1←DEN
[7]   'ENTER REMAINDER OF BRIDGE'
[8]   BRIDGE X←⎕
[9]   NUM←(NUM×D1)+DEN×N1
[10]  DEN←DEN×D1
[11]  ELT
[12]  VAL
    ∇
      ∇CYCLE[⎕]∇
    ∇ CYCLE A;M
[1]   M← 2 2 ρ 1 0 0 1
[2]   M←M+.× 2 2 ρA[1], 1 1 0
[3]   →2××A←1↓A
[4]   NUM←(-/ 1 1 ⍉⎕←M),1
[5]   DEN←M[2;1]×2
[6]   BASE←(NUM[1]*2)+4××/ 1 1 ⍉⌽M
[7]   ELT
    ∇
      ∇BRIDGE[⎕]∇
    ∇ BRIDGE X
[1]   INV
[2]   ADD ¯1↑X
[3]   →ιρX←¯1↓X
    ∇
      ∇ELT[⎕]∇
    ∇ ELT
[1]   '(';NUM[1];'+';NUM[2];'×';BASE;'*.5)÷';DEN
    ∇
      ∇VAL[⎕]∇
    ∇ E←VAL
[1]   E←+/(NUM×1,BASE*0.5)÷DEN
    ∇
      ∇INV[⎕]∇
    ∇ INV
[1]   DEN←(-/(1,BASE)×NUM×NUM)÷DEN
[2]   NUM← 1 ¯1 ×NUM
    ∇
```

```
      ∇ADD[⎕]∇
    ∇ ADD C
[1]   NUM←NUM+(DEN×C),0
    ∇
      M
ENTER CYCLE
⎕:
      2 1
  3  2
  1  1
(2+1×12*.5)÷2
ENTER FIRST BRIDGE
⎕:
      3 2 2
ENTER REMAINDER OF BRIDGE
⎕:
      0
(1.1032E5+530×12*.5)÷30264
3.7058


      M
ENTER CYCLE
⎕:
      3 2 2 3
  58  17
  17   5
(53+1×3965*.5)÷34
ENTER FIRST BRIDGE
⎕:
      0
ENTER REMAINDER OF BRIDGE
⎕:
      0
(0+¯68×3965*.5)÷¯1156
3.704


      M
ENTER CYCLE
⎕:
      1 2
  3  1
  2  1
(2+1×12*.5)÷4
ENTER FIRST BRIDGE
⎕:
      3 3 2 1
ENTER REMAINDER OF BRIDGE
⎕:
      0 2 2
(9.5389E7+93298×12*.5)÷2.5826E7
3.706
```

```
      ∇DIMENSION[⎕]∇
    ∇ DIMENSION
[1]   CYCLE 2 1
[2]   A←VAL
[3]   R←- 2 1 ρ(A-2),÷A
[4]   IL←((2 1 ρ1+÷A),R) CR 2 2 ρA- 1 0
[5]   IR←⌽((2 1 ρA),R) CR 2 2 ρ 2 1 +÷A
[6]   'BOUNDS ON SUBINTERVALS'
[7]   ⎕←II← 2 2 ρIL,IR
[8]   I←II[;1]
[9]   'LOWER BOUND ON DIMENSION'
[10]  NEWT 0 1
[11]  I←II[;2]
[12]  'UPPER BOUND ON DIMENSION'
[13]  NEWT 0 1
    ∇
      ∇CR[⎕]∇
    ∇ A←X CR Y
[1]   A←(×/X-Y)÷×/X-⌽Y
    ∇
      ∇NEWT[⎕]∇
    ∇ NEWT X
[1]   →1E¯10<|-/X←⎕←X[1]-(ORD X[1])×(÷DER X[1]),(-/X)÷-/ORD X
    ∇
      ∇ORD[⎕]∇
    ∇ Y←ORD X
[1]   Y←¯1++/[1] I∘.*X
    ∇
      ∇DER[⎕]∇
    ∇ Y1←DER X
[1]   Y1←+/(⍟I)×I*X
    ∇
      DIMENSION
  3  2
  1  1
(2+1×12*.5)÷2
BOUNDS ON SUBINTERVALS
    0.37669      0.39566
    0.16987      0.18146
LOWER BOUND ON DIMENSION
0.36376  0.68802
0.50361  0.53824
0.51957  0.51997
0.51975  0.51975
0.51975  0.51975
UPPER BOUND ON DIMENSION
0.37966  0.7028
0.52619  0.5608
0.54302  0.54342
0.54322  0.54322
0.54322  0.54322
```

```
      ∇TABLE2[⎕]∇
    ∇ TABLE2
[1]   CYCLE 2 1 1 1
[2]   D1←1+÷C1←1+÷B1←1+÷A1←VAL
[3]   CYCLE 2 2 2 1
[4]   D2←2+÷C2←2+÷B2←1+÷A2←VAL
[5]   PTS← 4 2 ρB1,A1,B2,C2,B1,A2,D1,C2
[6]   'POINTS:  ';PTS
[7]   RS←-÷ 4 2 ρC1,D1,A2,D2,(A2-1),B2,A1,D1+1
[8]   '-R:  ';RS
[9]   X←PTS[; 2 2 1 ρ 1 1 2 2],RS[; 2 2 1 ρ 2 1 1 2]
[10]  Y←(4 2 2 1 ρC1,C1,D2,D2),PTS[; 2 2 1 ρ 2 2 1 1]
[11]  V←X CR Y
[12]  TYPE← 4 2 ρ 1 4 3 2 1 2 1 2
[13]  I←⍉V[;;2]
[14]  'MEASURE IN DIMENSION .5'
[15]  Z←0<⎕←ORD 0.5
[16]  'ONLY ';Z/⍳4;' CAN GIVE TROUBLE'
    ∇
      TABLE2
  8  5
  3  2
(6+1×96*.5)÷6
  17  12
   7   5
(12+1×480*.5)÷14
POINTS:
     1.3798       2.633
     1.4129       2.7078
     1.3798       2.4221
     1.5798       2.7078
-R:
   ¯0.5798      ¯0.63299
   ¯0.41287     ¯0.42206
   ¯0.7032      ¯0.70778
   ¯0.3798      ¯0.38763
MEASURE IN DIMENSION .5
¯0.0049517  0.0074817  ¯0.16199  ¯0.10305
ONLY 2 CAN GIVE TROUBLE

      I
   0.38374     0.35161     0.42601     0.18853
   0.14106     0.17183     0.034341    0.21414

      TYPE[2;]
3  2
      I←I[2;2],I[1;2]×I[;3]
      ORD .5
¯0.088573
```

```
      ∇SECT4[⎕]∇
    ∇ SECT4
[1]   SΔSECT4←LOOP
[2]   CYCLE 1 3
[3]   C←(⍳3)+B←END←÷A←VAL
[4]   R← ¯2 ¯1 ,÷¯3
[5]   F2←F1← 3 1 ⍴⍳3
[6]   E1←(3 2 ⍴B,A),C[F1],R[F1]
[7]   E2←E1[; 3 2 1 4]
[8]   KEY←'ONLY [1;1] SHOULD BE 0'
[9]   COMP
[10]  E11←E1[,1; 1 2]
[11]  E22←E2[,1; 1 2]
[12]  J←,1
[13] LOOP:F1← 1 1 ⍴3
[14]  F2← 1 1 ⍴1
[15]  E1←(÷1+÷2+E11[J;]),A,R[F1]
[16]  E2←(E22[J;]+2),A,R[F2]+2
[17]  KEY←'SHOULD BE 1, THEN FACING CORNERS NOT CUT'
[18]  COMP
[19]  →GOOD/NEXT
[20]  E2[1;4]←R[F2←3]
[21]  'TRY FOR CONTAINED INTERVALS'
[22]  COMP
[23] NEXT:E1← 3 4 ⍴E1
[24]  E1[;3]←C[F1← 3 1 ⍴⍳3]
[25]  E2←(3 2 ⍴E2[; 1 2]),(3 1 ⍴END←END⌊E[1;1]),R[F2←F1]
[26]  KEY←'∨ WITH TRANSPOSE CHECKED BELOW'
[27]  COMP
[28]  GOOD∨⍉GOOD
[29]  'SHOULD BE 1, THEN OUTER CORNER NOT CUT'
[30]  F2←F1
[31]  E1[; 1 2]←÷E1[; 2 1]
[32]  E2←(1+÷E2[3⍴1; 2 1]),(3 1 ⍴B),R[F2]
[33]  KEY←'[I;J]=[J;I]=0 ONLY FOR [1;1]. CHECKED BELOW'
[34]  COMP
[35]  GOOD∨⍉GOOD
[36]  'ONLY [1;1] SHOULD BE 0'
[37]  E11←E11,[1] E1[,1; 1 2],[1] 1+÷1+÷2+E11[J;]
[38]  E22←E22,[1] E2[,1; 1 2],[1] 1+÷1+÷2+E22[J;]
[39]  J←J+1
[40]  →LOOP
    ∇

      ∇COMP[⎕]∇
    ∇ COMP
[1]   'TYPE:  ';,F2;F1
[2]   GOOD←⎕←(Y←E1[; 1 4] CR E1[; 2 3])∘.<X←E2[; 1 4] CR E2[;
      2 3]
[3]   KEY
[4]   'X=';X
[5]   'Y=';Y
    ∇
```

```
      SECT4
  4  1
  3  1
(3+1×21*.5)÷6
TYPE:  1  2  3
  1
  2
  3
 0 1 1
 1 1 1
 1 1 1
ONLY [1;1] SHOULD BE 0
X=0.45116  0.60436  0.59327
Y=0.54884  0.39564  0.40673

SECT4[13]
      →13
TYPE:  1
  3
 1
SHOULD BE 1, THEN FACING CORNERS NOT CUT
X=0.080816
Y=0.080532
TYPE:  1  2  3
  1
  2
  3
 1 1 0
 1 1 1
 1 1 1
∨ WITH TRANSPOSE CHECKED BELOW
X=0.08976  0.070314  0.051343
Y=0.053566  0.040444  0.035913
 1 1 1
 1 1 1
 1 1 1
SHOULD BE 1, THEN OUTER CORNER NOT CUT
TYPE:  1  2  3
  1
  2
  3
 0 0 0
 1 1 0
 1 1 1
[I;J]=[J;I]=0 ONLY FOR [1;1]. CHECKED BELOW
X=0.070783  0.06549  0.058279
Y=0.13434  0.064515  0.05089
 0 1 1
 1 1 1
 1 1 1
ONLY [1;1] SHOULD BE 0

SECT4[13]
```

```
      →13
TYPE:  1
  3
 1
SHOULD BE 1, THEN FACING CORNERS NOT CUT
X=0.0080816
Y=0.0080608
TYPE:  1  2  3
  1
  2
  3
 1 1 0
 1 1 1
 1 1 1
∨ WITH TRANSPOSE CHECKED BELOW
X=0.0086225  0.0067545  0.0049321
Y=0.0051964  0.0038654  0.0034149
 1 1 1
 1 1 1
 1 1 1
SHOULD BE 1, THEN OUTER CORNER NOT CUT
TYPE:  1  2  3
  1
  2
  3
 0 0 0
 1 0 0
 1 1 1
[I;J]=[J;I]=0 ONLY FOR [1;1]. CHECKED BELOW
X=0.0064824  0.0059668  0.0052729
Y=0.012293  0.0059738  0.0047232
 0 1 1
 1 0 1
 1 1 1
ONLY [1;1] SHOULD BE 0

SECT4[13]
      E1[;1]←3ρB
      E2[;1]←C
      COMP
TYPE:  1  2  3
  1
  2
  3
 0 0 0
 0 1 1
 0 1 1
[I;J]=[J;I]=0 ONLY FOR [1;1]. CHECKED BELOW
X=0.41318  0.57916  0.57037
Y=0.66391  0.4882  0.42963
```

THIS WAS A SPECIAL CASE; ONLY [2;2] IS OF INTEREST.

```
      →13
TYPE:  1
  3
 0
SHOULD BE 1, THEN FACING CORNERS NOT CUT
X=0.0034289
Y=0.0038763
TRY FOR CONTAINED INTERVALS
TYPE:  3
  3
 1
SHOULD BE 1, THEN FACING CORNERS NOT CUT
X=0.0039809
Y=0.0038763
TYPE:  1  2  3
  1
  2
  3
 1 1 0
 1 1 1
 1 1 1
∨ WITH TRANSPOSE CHECKED BELOW
X=0.0040114  0.0030795  0.0022056
Y=0.002443  0.0017986  0.0015834
 1 1 1
 1 1 1
 1 1 1
SHOULD BE 1, THEN OUTER CORNER NOT CUT
TYPE:  1  2  3
  1
  2
  3
 0 0 0
 1 1 1
 1 1 1
[I;J]=[J;I]=0 ONLY FOR [1;1]. CHECKED BELOW
X=0.0032617  0.0030178  0.0026855
Y=0.0052443  0.0026576  0.0021191
 0 1 1
 1 1 1
 1 1 1
ONLY [1;1] SHOULD BE 0

SECT4[13]
```

```
      →13
TYPE:  1
  3
 1
SHOULD BE 1, THEN FACING CORNERS NOT CUT
X=0.00074012
Y=0.00073814
TYPE:  1  2  3
  1
  2
  3
 1 1 0
 1 1 1
 1 1 1
∨ WITH TRANSPOSE CHECKED BELOW
X=0.0007932  0.0006203  0.00045219
Y=0.00047607  0.00035422  0.00031296
 1 1 1
 1 1 1
 1 1 1
SHOULD BE 1, THEN OUTER CORNER NOT CUT
TYPE:  1 ¯2  3
  1
  2
  3
 0 0 0
 1 1 0
 1 1 1
[I;J]=[J;I]=0 ONLY FOR [1;1]. CHECKED BELOW
X=0.00059784  0.00055029  0.00048629
Y=0.0011398  0.00055011  0.00043436
 0 1 1
 1 1 1
 1 1 1
ONLY [1;1] SHOULD BE 0

SECT4[13]
     E11                                E22
    0.79129      1.2638                 1.7913      1.2638
    1.3064       1.3583                 1.3064      1.2638
    1.7362       1.7655                 1.7913      1.7655
    1.2978       1.3024                 1.3064      1.3024
    1.7678       1.7706                 1.7678      1.7655
    1.2656       1.2676                 1.2656      1.2638
    1.7889       1.7902                 1.7913      1.7902
    1.3028       1.3032                 1.3028      1.3024
    1.7673       1.7676                 1.7678      1.7676

      )OFF
002  16.57.58 11/09/72 BUM
CONNECTED     0.24.23  TO DATE     0.42.47
CPU TIME      0.00.05  TO DATE     0.00.08
WWN
```

DISTINCT PRIME FACTORS OF CONSECUTIVE INTEGERS

P.L. Cijsouw and R. Tijdeman

1. C.A. Grimm [4] stated the following conjecture:
Let $n+1,\ldots,n+k$ be consecutive composite integers. Then there exist k distinct primes $p_1,\ldots,p_k$ such that $p_j|(n+j)$ for $j = 1,\ldots,k$.
Let $k_n (n = 1,2,\ldots)$ be the smallest integer such that no distinct primes $p_1,\ldots,p_{k_n}$ exist subject to $p_j|(n+j)$ for $j = 1,\ldots,k_n$, for example $k_{23}=7$.
Grimm's conjecture is equivalent to the statement:
For every positive integer n there is a prime p with $n < p < n+k_n$.
The application of Hall's Theorem on distinct representatives yields the following characterization of k_n:

k_n is the smallest integer for which integers $n < n_1 < n_2 < \ldots < n_v \leq n+k_n$ can be found such that the number of distinct prime factors of $\Pi_{j=1}^{v} n_j$ is less then v.
Both the relation with Grimm's conjecture and this characterization make the sequence $\{k_n\}_{n=1}^{\infty}$ very interesting.

Grimm himself proved $k_n^{k_n-1} > n$ for all n.

Hence, $k_n > \log n/\log\log n$. Erdős and Selfridge [2, Theorem 3] improved upon this by the inequality $k_n > (1 + o(1))\log n$. In view of [1] this result shows that Grimm's conjecture holds for infinitely many sets of composite numbers between consecutive primes. They conjectured $k_n/(\log n)^{\alpha} \to \infty$ for every $\alpha > 0$. On the other hand, one of their results implies $k_n < c_1(n/\log n)^{1/2}$ for some absolute constant c_1. In particular, even a weaker form of Grimm's conjecture would imply that there are primes between n^2 and $(n+1)^2$ for all sufficiently large n. Ramachandra [6] used Baker's method on linear forms in the logarithms of algebraic numbers and proved

$$k_n \gg \log n(\log\log n/\log\log\log n)^{1/2} .$$

We prove here

THEOREM 1. There exists an effectively computable absolute constant C such that

$$k_n \geq C(\log n)^2/(\log\log n)^6 \quad \text{for} \quad n > 1.$$

The proof of this theorem goes via two results which are of interest by themselves. The first is one on diophantine approximations:

THEOREM 2. Let $a, a_1, a_2, b, b_1, b_2, L$ be positive integers subject to $\max(a_1, a_2) \leq a$, $\max(b_1, b_2) \leq b$, $b_1 \neq b_2$ and

$$(1) \quad 10^4(1 + \log a)(\log b + \log L)^4 \leq L .$$

Then there exists an effectively computable

absolute constant S_0 such that

$$(2) \quad \left|\log \frac{a_1}{a_2} + L \log \frac{b_1}{b_2}\right| > \exp\{-10^9(1 + \log a)(\log b + \log L)^4\} ,$$

if $bL > S_0$.

The proof of Theorem 2 uses a variant of Gel'fond's method on transcendence measures. See [3,Ch. III,. §5]. We remark that a similar theorem can be stated with arbitrary algebraic numbers α and β instead of rational numbers a_1/a_2 and b_1/b_2. Very recently Shorey proved a result which is closely related to Theorem 2 (see section 4). The application of his result would lead to a certain modification of the following theorem, which states that an interval cannot contain "too many" integers composed of small prime factors and which is deduced from Theorem 2:

<u>THEOREM 3.</u> Let $\delta > 0$, $0 < \theta < 1$. Let $p_1,\dots,p_r$ be distinct primes not exceeding m. Let $n,k,n_1,\dots,n_t$ be integers such that

$$0 < n < n_1 < n_2 < \dots < n_t \leqslant n+k \leqslant n+n^{1-\theta} ,$$

and such that, moreover, $\Pi_{j=1}^{t} n_j$ has no other prime factors than $p_1,\dots,p_r$. Suppose

$$(3) \qquad r \leqslant (2 - \delta)t$$

and

$$(4) \qquad \theta \log n \geqslant 10^{13}(\log m)^4 .$$

Then there exist effectively computable absolute constants C_1 and n_0 such that

$$(5)\quad t \leqslant \frac{C_1}{\delta\theta}\left(\frac{k \log k}{\log n} + \log m\right)(\log m + \log\log n)^4$$

for $n \geqslant n_0$.

The proof of Theorem 1 is completed by an application of Hall's Theorem on distinct representatives and an argument also used by Erdös and Selfridge [2].

2. In the proof of Theorem 2 the following two lemmas are used:

<u>LEMMA 1.</u> The system of r equations in s variables

$$\sum_{\sigma=1}^{s} A_{\rho\sigma}\, x_\sigma = 0 \quad (\rho = 1,\ldots,r)$$

with $s > r$, $A_{\rho\sigma} \in \mathbb{Z}$, $|A_{\rho\sigma}| \leqslant A$ for $\rho = 1,\ldots,r$ and $\sigma = 1,\ldots,s$ has a nontrivial solution $x_\sigma = C_\sigma (\sigma = 1,\ldots,s)$ with $C_\sigma \in \mathbb{Z}$ and $|C_\sigma| \leqslant [(sA)^{r/(s-r)}] + 2$ for $\sigma = 1,\ldots,s$.

<u>Proof</u>. See e.g. [7, Hilfssatz 29].

<u>LEMMA 2.</u> Let K, T^* and P be positive integers. Let

$$F(z) = \sum_{k=0}^{K-1} C_k e^{\omega_k z},$$

with $C_k \in \mathbb{C}$, $\omega_k \in \mathbb{R}$ and $\omega_k \neq \omega_j$ when $k \neq j$ $(k, j = 0,1,\ldots,K-1)$. Put $\Omega = \max(1, \max_k |\omega_k|)$ and $\omega = \min(1, \min_{k\neq j} |\omega_k - \omega_j|)$.

If $T^*P \geqslant 2K + 13\Omega P$, then

$$\max_k |C_k| \leqslant P\left(\frac{1}{3\omega\max(1,P-1)}\right)^{K-1} e^{7\Omega P}\ 72^{T^*P} \times$$

$$\max_{\substack{t=0,\dots,T^*-1 \\ p=0,\dots,P-1}} |F^{(t)}(p)| .$$

Proof. See [2, Theorem 3 (ii),(iii)].

Proof of Theorem 2. Put $S = \log b + \log L$. We shall frequently use the inequalities

$$(6)\quad \log(1 + \log a) \leqslant S, \quad |\log\frac{b_1}{b_2}| \geqslant e^{-2S} .$$

The first is a direct consequence of (1). The second is trivial for $b_1/b_2 > 2$, while for $b_1/b_2 \leqslant 2$ it follows from

$$|\log \frac{b_1}{b_2}| \geqslant |\frac{b_1}{b_2} - 1| \cdot \frac{1}{2} \geqslant \frac{1}{2b} \geqslant e^{-2S} .$$

Suppose that

$$(7)\quad |\log \frac{a_1}{a_2} + L \log \frac{b_1}{b_2}| \leqslant \exp\{-10^9 S^4(1+\log a)\} ,$$

for a set of positive integers a,a_1,a_2,b,b_1,b_2,L. We shall derive a finite number of estimates which are valid for sufficiently large S. S_0 can be chosen as the maximum of the bounds thus obtained. Eventually we shall obtain a contradiction for $S \geqslant S_0$.

Put $K = [10^4 S(1+\log a)]$, $M = [10^4 S^2]$,

$P = [10^2 S]$, $T = [4.10^5 S^2(1+\log a)]$, $T^* = [25.10^5 S^2(1+\log a)]$. We consider the auxiliary function

$$F(z) = \sum_{k=0}^{K-1} \sum_{m=0}^{M-1} C_{km}\, e^{(k-mL)(\log \frac{b_1}{b_2})z} ,$$

where the coefficients C_{km} will be chosen later. For $t = 0,1,2,\ldots$ and $p = 0,1,2,\ldots$ we define

$$F_{tp} = (\log \frac{b_1}{b_2})^{-t} F^{(t)}(p)$$

$$= \sum_k \sum_m C_{km}(k-mL)^t \left(\frac{b_1}{b_2}\right)^{(k-mL)p}$$

and

$$\Phi_{tp} = \sum_k \sum_m C_{km}(k-mL)^t \left(\frac{b_1}{b_2}\right)^{kp} \left(\frac{a_1}{a_2}\right)^{mp} .$$

We have, for $t = 0,1,\ldots,T-1$ and $p = 0,1,\ldots,P-1$,

$$b_2^{KP}\, a_2^{MP}\, \Phi_{tp} = \sum_k \sum_m C_{km}\, A_{tpkm} ,$$

where, by (6), the integers A_{tpkm} satisfy $|A_{tpkm}| \leq \exp\{6.10^6 S^3(1+\log a)\}$. We know $KM \geq 2\,TP$. According to Lemma 1 we choose the coefficients C_{km} as integers, not all zero, such that

$$|C_{km}| \leq \exp\{7.10^6 S^3(1+\log a)\} \quad \text{for}$$

$$k = 0,1,\ldots,K-1 \text{ and } m = 0,1,\ldots,M-1 ,$$

and such that, moreover,

(8) $\Phi_{tp} = 0$ for $t = 0,1,\ldots,T-1$ and

$p = 0,1,\ldots,P-1$.

For $t = 0,1,\ldots,T^*-1$ and $p = 0,1,\ldots,P-1$ the difference $|\Phi_{tp}-F_{tp}|$ can be estimated as follows:

We have from (7)

$$\left|\frac{a_1}{a_2} - \left(\frac{b_1}{b_2}\right)^{-L}\right| = \left|\exp(\log \frac{a_1}{a_2}) - \exp(-L \log \frac{b_1}{b_2})\right|$$

$$\leq \left|\log \frac{a_1}{a_2} + L \log \frac{b_1}{b_2}\right| \exp(\left|\log \frac{a_1}{a_2}\right| + 1)$$

$$\leq \exp\{-9.10^8 S^4(1+\log a)\} .$$

Thus

$$\left| \left(\frac{a_1}{a_2}\right)^{mp} - \left(\frac{b_1}{b_2}\right)^{-Lmp} \right| \leq MP\left|\frac{a_1}{a_2} - \left(\frac{b_1}{b_2}\right)^{-L}\right| \left(\frac{a_1}{a_2} + 1\right)^{MP-1}$$

$$\leq \exp\{-8.10^8 S^4(1+\log a)\} ,$$

whence

$$(9) \qquad |\Phi_{tp}-F_{tp}|$$

$$\leq KM \max_{k,m}|C_{km}|(2KML)^{T^*} b^{KP} \exp\{-8.10^8 S^4(1+\log a)\}$$

$$\leq \exp\{-7.10^8 S^4(1+\log a)\} \text{ for } t = 0,1,\ldots,T^*-1$$

$$\text{and } p = 0,1,\ldots,P-1.$$

We see from (8) that $|F_{tp}| \leq \exp\{-7.10^8 S^4(1+\log a)\}$ for $t = 0,1,\ldots,T-1$ and $p = 0,1,\ldots,P-1$.

Hence, for these values of t and p ,

$$|F^{(t)}(p)| \leq (\log b)^T |F_{tp}|$$
$$\leq \exp\{-6.10^8 S^4 (1 + \log a)\} .$$

In view of (7)

$$|(k-mL)\log \frac{b_1}{b_2}| \leq k \log b + m(1+\log a)$$
$$\leq 2.10^4 S^2 (1+\log a),$$

for $k = 0,1,\ldots,K-1$ and $m = 0,1,\ldots,M-1$. Hence,

$$\max_{|z|=10P} |F(z)| \leq \exp\{28.10^6 S^3 (1 + \log a)\} .$$

We use the following well known consequence of Cauchy's residue theorem: If $|z| = 2P$, then

$$F(z) = \frac{1}{2\pi i} \int_{|\zeta|=10P} \frac{F(\zeta)}{\zeta - z} \prod_{p=0}^{P-1} \left(\frac{z-p}{\zeta-p}\right)^T d\zeta +$$
$$- \sum_{p=0}^{P-1} \sum_{t=0}^{T-1} \frac{F^{(t)}(p)}{t!} \frac{1}{2\pi i} \int_{|\zeta-p|=\frac{1}{2}} \frac{(\zeta-p)^t}{\zeta-z} \prod_{q=0}^{P-1} \left(\frac{z-q}{\zeta-q}\right)^T d\zeta .$$

It follows that

$$\max_{|z|=2P} |F(z)| \leq 10P \max_{|z|=10P} |F(z)| \cdot \left(\frac{3P}{9P}\right)^{PT}$$
$$+ PT \max_{\substack{t=0,1,\ldots,T-1 \\ p=0,1,\ldots,P-1}} |F^{(t)}(p)| \cdot \left(\frac{3P}{1/2}\right)^{PT}$$
$$\leq \exp\{-11.10^6 S^3 (1+\log a)\} + \exp\{-5.10^8 S^4 (1+\log a)\}$$
$$\leq 2 \exp\{-11.10^6 S^3 (1+\log a)\} .$$

For $t = 0,1,\ldots,T^*-1$ and $p = 0,1,\ldots,P-1$ we obtain

$$|F^{(t)}(p)| = \left|\frac{t!}{2\pi i}\int_{|z|=2P}\frac{F(z)}{(z-p)^{t+1}}\,dz\right|$$

$$\leqslant T^{*T^*}\,2P\max_{|z|=2P}|F(z)| \leqslant \exp\{-8.10^6 S^3(1+\log a)\}\ .$$

Hence,

$$|F_{tp}| = \left|\left(\log\frac{b_1}{b_2}\right)^{-t}F^{(t)}(p)\right| \leqslant e^{2ST^*}|F^{(t)}(p)|$$

$$\leqslant \exp\{-3.10^6 S^3(1+\log a)\}\ .$$

We see from (9) that

$$|\Phi_{tp}| < \exp\{-2.10^6 S^3(1+\log a)\}$$

for $t = 0,1,\ldots,T^*-1$ and $p = 0,1,\ldots,P-1$. However, Φ_{tp} is a rational number with denominator at most

$$b_2^{KP}\,a_2^{MP} \leqslant \exp\{2.10^6 S^3(1+\log a)\}\ .$$

Consequently $\Phi_{tp} = 0$ for $t = 0,1,\ldots,T^*-1$ and $p = 0,1,\ldots,P-1$. Hence, from (9),

$$|F_{tp}| \leqslant \exp\{-7.10^8 S^4(1+\log a)\}$$

and

$$|F^{(t)}(p)| \leqslant \exp\{-6.10^8 S^4(1+\log a)\}$$

for $t = 0,1,\ldots,T^*-1$ and $p = 0,1,\ldots,P-1$.

We now apply Lemma 2. From (1) we see that $K \leqslant L$. The numbers $(k-mL)\log(b_1/b_2)$ in the exponents of F are therefore distinct, and, from (6),

$$\omega = \min_{(k_1,m_1)\neq(k_2,m_2)} |\{(k_1-m_1L)-(k_2-m_2L)\}\log \frac{b_1}{b_2}|$$

$$\geq |\log \frac{b_1}{b_2}| \geq e^{-2S} .$$

Furthermore, as we saw before,

$$\Omega = \max_{k,m} |(k-mL)\log \frac{b_1}{b_2}| \leq 2.10^4 S^2(1 + \log a) .$$

The inequality $PT^* \geq 2\ KM + 13\Omega P$ is easily checked. By the application of Lemma 2 to F, with KM instead of K, we obtain

$$\max_{k,m} |C_{km}| \leq \exp\{-10^8 S^4(1 + \log a)\} .$$

Thus $C_{km} = 0$ for $k = 0,1,\ldots,K-1$ and $m = 0,1,\ldots,M-1$, in contradiction with the choice of these integers.

3. Proof of Theorem 3. Without loss of generality we may assume that the following three inequalities hold.

$$(10)\begin{cases} \delta t > 4k \log k(\log\log n)^4/\log n, \\ \delta t > 20, \\ \delta t > 10^6(\frac{k \log k}{\log n}+\log m)(\log m+\log\log n)^2 . \end{cases}$$

In the sequel we denote the number of elements of a set A by $|A|$.

Put $P = \{p_1,\ldots,p_r\}$ and $N = \{n_1,\ldots,n_t\}$.

For each $p \in P$ we choose an element $f(p)$ from N such that no other element in N contains p to a higher order than $f(p)$ does. This induces a map $f:P \to N$. Let N_0, N_1, N_2 consist of the integers in N which are not in the range of f, the images of precisely one element in P, the images of at least two elements in P, respectively. We have $|N_0|+|N_1|+|N_2| = t$ and $|N_1| + 2|N_2| \leq r$. Hence

$$(11) \qquad 2|N_0| + |N_1| \geq 2t - r .$$

Define α_p by $p^{\alpha_p} | f(p)$, $p^{\alpha_p+1} \nmid f(p)$. The number of factors p in $\Pi_{\nu \in N} \nu$ does not exceed

$$\alpha_p + \left[\frac{k}{p}\right] + \left[\frac{k}{p^2}\right] + \left[\frac{k}{p^3}\right] + \cdots .$$

Since $\Pi_{\nu \in N} \nu$ has only prime factors from P, while

$$\prod_p p^{\left[\frac{k}{p}\right] + \left[\frac{k}{p^2}\right] + \left[\frac{k}{p^3}\right] + \cdots} = k! ,$$

we obtain

$$(12) \left(\prod_{\nu \in N_0} \nu\right)\left(\prod_{\substack{p \in P \\ f(p) \in N_1}} \frac{f(p)}{p^{\alpha_p}}\right) \leq \frac{\prod_{\nu \in N} \nu}{\prod_{p \in P} p^{\alpha_p}} \leq k! \leq k^k .$$

It follows that $n^{|N_0|} \leq k^k$. Hence, $|N_0| \leq k \log k / \log n$. This implies, by (11), (3) and (10), that

$$|N_1| \geq 2t - r - 2|N_0| \geq \frac{1}{2}\delta t .$$

The set N_1 consists of integers of the form $\nu = m_\nu p_\nu^{L_\nu}$, where

$\nu = f(p_\nu), L_\nu = \alpha_{p_\nu}, p_{\nu_1} \neq p_{\nu_2}$ for $\nu_1, \nu_2 \in N_1, \nu_1 \neq \nu_2$.

Furthermore, from (12),

$$\prod_{\nu \in N_1} m_\nu = \prod_{\substack{p \in P \\ f(p) \in N_1}} \frac{f(p)}{p^{\alpha_p}} \leq k^k .$$

Thus it is impossible that $m_\nu > \exp(4k\log k/\delta t)$ holds for more than $\delta t/4$ elements in N_1. There therefore exists a subset N_3 of N_1, consisting of at least $\delta t/4$ elements ν in N_1, such that, by (10),

$$m_\nu \leq \exp\left(\frac{4k\log k}{\delta t}\right) . \tag{13}$$

Hence

$$m_\nu \leq \exp(\log n/(\log\log n)^4) .$$

We have for $\nu \in N_3$ and n sufficiently large

$$L_\nu \leq \frac{\log(n+k)}{\log 2} \leq 2 \log n \tag{14}$$

and

$$L_\nu = \frac{\log(\nu/m_\nu)}{\log p_\nu} \geq \frac{\log n - \log n/(\log\log n)^4}{\log m} \geq \frac{\log n}{2 \log m} . \tag{15}$$

In view of the inequalities (14), $|N_3| > \delta t/4$ and (10) there exist distinct elements

$\nu_1, \nu_2 \in N_3$ such that

$$(16) \qquad |L_{\nu_1} - L_{\nu_2}| \leqslant \frac{2 \log n}{|N_3|-1} \leqslant \frac{10 \log n}{\delta t} .$$

We may assume $L_{\nu_1} \leqslant L_{\nu_2}$. We apply Theorem 2 with

$$L = L_{\nu_1}, \; a_1 = m_{\nu_1}, \; a_2 = m_{\nu_2} p_{\nu_2}^{L_{\nu_2} - L_{\nu_1}}, \; b_1 = p_{\nu_1},$$

$$b_2 = p_{\nu_2} ,$$

$$a = [\exp\{\frac{4k \log k}{\delta t} + \frac{10 \log m \log n}{\delta t}\}] \text{ and } b = m .$$

We have $b_1 \neq b_2$ and $\max(b_1, b_2) \leqslant b$. Furthermore, we see from (13) and (16) that $\max(a_1, a_2) \leqslant a$. It follows from (4) that, putting $Z = \log m + 2 \operatorname{loglog} n$,

$$\begin{aligned} (17) \qquad 10^{11} Z^4 &= 10^{11}(\log m + 2 \operatorname{loglog} n)^4 \\ &\leqslant 10^{13} \max((\log m)^4, (\operatorname{loglog} n)^4) \\ &< \theta \log n, \text{ for } n \text{ large.} \end{aligned}$$

Hence, by (10) and (15),

$$\begin{aligned} & 10^4(1 + \log a)(\log b + \log L) \\ \leqslant \; & 10^4 Z(1 + \frac{10(k \log k + \log m \log n)}{\delta t}) \\ \leqslant \; & \frac{\log n}{10 Z} + \frac{\log n}{10 Z} \leqslant \frac{\log n}{2 \log m} \leqslant L . \end{aligned}$$

Finally, $bL \geqslant m \log n/(2 \log m) \geqslant S_0$ for sufficiently large n. All conditions of Theorem 2 have been checked. It follows from (2) that

$$|\log \frac{\nu_1}{\nu_2}| = \left| \log \frac{m_{\nu_1}}{m_{\nu_2} p_{\nu_2}^{L_{\nu_2} - L_{\nu_1}}} + L_{\nu_1} \log \frac{p_{\nu_1}}{p_{\nu_2}} \right|$$

$$> \exp\{-10^{10}(1+ \frac{k \log k+\log m \log n}{\delta t})Z^4\} .$$

On the other hand, using $k < n^{1-\theta}$ we have

$$|\log \frac{\nu_1}{\nu_2}| \leqslant \log \frac{n+k}{n} < \frac{k}{n} < \exp(-\theta \log n) .$$

This implies

$$\theta \log n \leqslant 10^{10} Z^4 + \frac{10^{10} Z^4}{\delta t}(k \log k + \log m \log n) .$$

We see from (17) that $10^{10} Z^4 \leqslant \theta \log n/10$. Hence,

$$t \leqslant \frac{10^{11} Z^4}{9 \delta \theta}(\frac{k \log k}{\log n} + \log m) .$$

This proves the theorem.

4. Several variations of Theorem 3 are possible. For example, if a nontrivial lower bound for $\min_j p_j$ is given, then inequality (14) can be improved upon. If $|N_3|$ exceeds the difference between the right hand sides of (14) and (15), then $L_{\nu_1} = L_{\nu_2}$ for two different elements ν_1 and ν_2 in N_3. This leads to an improvement in formula (4). One can also replace Theorem 2 by similar results, e.g. by a recent theorem of T.N. Shorey [8, Theorem 1] or by a more general result

of K. Ramachandra [6]. Using the latter result Ramachandra [6] proved

$$k_n \gg \log n(\log\log n)^{1/2}(\log\log\log n)^{-1/2} .$$

We shall improve upon this by Theorem 1. This is achieved by using a more refined combinatorial argument, which is essentially due to Erdös and Selfridge [2].

5. For the proof of Theorem 1 we need Hall's Theorem on distinct representatives:

<u>LEMMA 3. (P. Hall, 1935)</u>. Let $S_1,\ldots,S_k$ be arbitrary sets. Then the nonexistence of distinct elements $\pi_1,\ldots,\pi_k$ with $\pi_j \in S_j$ for $j = 1,\ldots,k$ is equivalent to the existence of integers $0 < \rho_1 < \rho_2 < \ldots < \rho_t \leqslant k$ such that $\bigcup_{j=1}^{t} S_{\rho_j}$ contains less than t elements.

<u>Proof</u>. See e.g. [5, 45-47].

<u>Proof of Theorem 1.</u> Let n be a fixed positive integer such that $k_n \leqslant (\log n)^2$. Put $k = k_n$. Let S_j be the set of prime factors of $n + j$ for $j = 1,2,\ldots$. According to its definition k is the smallest integer such that there are no distinct representatives $\pi_j \in S_j$ for $j = 1,\ldots,k$. Hence, by Lemma 3, k is the smallest integer for which there exist integers $n < n_1 < n_2 < \ldots < n_t \leqslant n+k$ such that the number of distinct prime factors of $\prod_{j=1}^{t} n_j$ is less than t. Without loss of generality we may assume that t is minimal, i.e. that there is no set of v integers in

$(n, n+k]$ with $v < t$ such that the number of distinct prime factors of the product of these v integers is less than v. Denote the set of prime factors of $\Pi_{j=1}^{t} n_j$ by $P = \{p_1,\ldots,p_r\}$. Hence $r < t$. In consequence of the minimality of t, every prime $p \in P$ divides at least two elements of $N = \{n_1,\ldots,n_t\}$. Hence,

$$(18) \qquad p < k \qquad \text{for } p \in P.$$

We apply Theorem 3 with $\delta = 1$, $\theta = \frac{1}{2}$, $m = k$. It follows that there exist effectively computable absolute constants C_2 and n_0 such that

$$(19) \qquad t < C_2\left(\frac{k \log k}{\log n} + \operatorname{loglog} n\right)(\operatorname{loglog} n)^4$$

$$\text{for } n \geq n_0 .$$

The following argument due to Erdős and Selfridge gives a lower bound for t. Let $f\colon P \to N$ be defined as in the proof of Theorem 3. We know $|P| < |N|$. Let $n^* = \Pi_{p\in P} p^{\beta_p}$ be an element of N which is not in the range of f. Since $p^{\beta_p} | f(p)$ and $p^{\beta_p} | n^*$, we have $|f(p) - n^*| < k$, and hence

$$n < n^* = \Pi_{p\in P}\, p^{\beta_p} < k^{|P|} \leq k^t .$$

It follows that

$$(20) \qquad t > \frac{\log n}{\log k} .$$

We see from (19), (20) and the inequality $k < (\log n)^2$ that

$$\log n < 4C_2 \frac{k(\log\log n)^6}{\log n} + 2C_2(\log\log n)^6 .$$

Since $2C_2(\log\log n)^6 < \frac{1}{2} \log n$ for n sufficiently large, we obtain

$$(21) \qquad k > \frac{1}{8C_2} \frac{(\log n)^2}{(\log\log n)^6} \text{ for } n \geq n_0^* ,$$

where n_0^* is an effectively computable constant. It is obvious that there exists an effectively computable constant C_3 such that

$$(22) \qquad k > C_3 \frac{(\log n)^2}{(\log\log n)^6} \quad \text{for} \quad 1 < n < n_0^* .$$

The inequalities (21) and (22) prove Theorem 1.

REFERENCES

[1] P. Erdős, The difference of consecutive prime numbers, Duke Math. J., 6 (1940), 438-441.

[2] P. Erdős and J.L. Selfridge, Some problems on the prime factors of consecutive integers II, Proc. Wash. State Univ. Conference on Number Theory, Pullman (Wash.), 1971.

[3] A.O. Gel'fond, Transcendental and algebraic numbers, Dover, 1960.

[4] C.A. Grimm, A conjecture on consecutive composite numbers, Amer. Math. Monthly, 76 (1969), 1126-1128.

[5] M. Hall, Combinatorial theory, Blaisdell, 1969.

[6] K. Ramachandra, Application of Baker's theory to two problems considered by Erdős and Selfridge, J. Indian Math. Soc., (to appear).

[7] Th. Schneider, Einführung in die transzendenten Zahlen, Springer, 1957.

[8] T.N. Shorey, Linear forms in the logarithms of algebraic numbers with small coefficients

I, to appear.

[9] R. Tijdeman, An auxiliary result in the theory of transcendental numbers, to appear in J. Number Theory.

Instituut voor Propedeutische Wiskunde,
Univ. of Amsterdam, Amsterdam, Netherlands

Mathematisch Instituut, R. U. Leiden,
Leiden, Netherlands

LINEAR RELATIONS BETWEEN $2\pi i$ AND THE PERIODS OF TWO ELLIPTIC CURVES

John Coates

I. Introduction.

Let n be any positive integer. Suppose that we are given n elliptic curves

$$(1) \qquad y^2 = 4x^3 - g_{2,k}x - g_{3,k} \quad (1 \leqslant k \leqslant n) ,$$

whose coefficients $g_{2,k}, g_{3,k}$ $(1 \leqslant k \leqslant n)$ are algebraic numbers. Let $\wp_k(z)$ be the $\wp$-function with invariants $g_{2,k}, g_{3,k}$, and let $\zeta_k(z)$ be the associated Weierstrass ζ-function. Choose a basis $\omega_{1,k}, \omega_{2,k}$ of the period lattice of $\wp_k(z)$ with $I(\omega_{1,k}/\omega_{2,k}) > 0$, and, as usual, let $\eta_{j,k} = 2\zeta_k(\omega_{j,k}/2)$ $(j=1,2)$. By theorems of Lindemann and Schneider, each of the numbers

$$(2) \quad \omega_{1,k}, \; \omega_{2,k}, \; \eta_{1,k}, \; \eta_{2,k} \quad (1 \leqslant k \leqslant n), \; 2\pi i$$

is transcendental. Moreover, these numbers are particularly interesting from the point of view of the theory of transcendental numbers because the following algebraic relations exist between them. For $k = 1,\ldots,n$, we have the Legendre relation

$$\omega_{1,k}\,\eta_{2,k} - \omega_{2,k}\,\eta_{1,k} = 2\pi i .$$

In addition, the numbers $\omega_{1,k}$, $\omega_{2,k}$, $\eta_{1,k}$, $\eta_{2,k}$ $(1 \leq k \leq n)$ may well be linearly dependent over the rational field $\mathbb{Q}$ if $n > 1$. Also, if the elliptic curve (1) has complex multiplications, then $\omega_{1,k}/\omega_{2,k}$ is an imaginary quadratic irrationality. Finally, in the special case in which either $g_{2,k} = 0$ or $g_{3,k} = 0$ (whence, of course, (1) does have complex multiplications), it is easily seen that we also have $\eta_{2,k} = (\omega_{1,k}/\omega_{2,k})\eta_{1,k}$. These seem to be the only known algebraic relations between the numbers (2). Indeed, it is probable that these relations, together with their consequences, exhaust all polynomial relations with algebraic coefficients between the numbers (2). Needless to say, a proof of this assertion in general seems to be quite beyond the present resources of the theory of transcendental numbers. However, it does seem plausible to study this problem for the special case of linear relations. When $n = 1$, much of this has already been done by Schneider [6], Baker [1],[2], and the author [3],[4]. Schneider and Baker's results also include particular cases of the above conjecture for $n = 2$, and one might suspect that the methods used for $n = 1$ would in general be applicable for $n \geq 2$. Unfortunately, this is definitely not the case, and new ideas seem to be needed to settle the problem even for two elliptic curves. However, in the present paper we show that the old methods do give results for two elliptic curves when at least one of the two curves has complex multiplications.

THEOREM 1. Assume that $n = 2$, and that both the curves (1) have complex multiplications. Then any nonvanishing linear form in $\omega_{1,k}$, $\omega_{2,k}$, $\eta_{1,k}$, $\eta_{2,k}$ $(k=1,2)$, $2\pi i$, with algebraic coefficients, is transcendental.

THEOREM 2. Assume that $n = 2$, and that one of the curves (1), say that given by $k = 1$, has complex multiplications. Then any nonvanishing linear form in $2\pi i$, $\omega_{1,1}$, $\omega_{2,1}$, $\eta_{1,1}$, $\eta_{2,1}$, and $\omega_{1,2}$, $\eta_{1,2}$ (or equally well $\omega_{2,2}$, $\eta_{2,2}$ instead of $\omega_{1,2}$, $\eta_{1,2}$), with algebraic coefficients, is transcendental.

A particular case of Theorem 1 may be worth mentioning. The two elliptic curves

$$y^2 = 4x^3 - 4x, \quad y^2 = 4x^3 - 4$$

both have complex multiplications. A period of each of the curves is given respectively by

$$2\int_1^\infty \frac{dx}{\sqrt{4x^3-4x}} = \frac{\Gamma(1/4)^2}{(8\pi)^{1/2}}, \quad 2\int_1^\infty \frac{dx}{\sqrt{4x^3-4}} = \frac{\Gamma(1/3)^3}{\pi 2^{8/3}}$$

Hence, for all algebraic numbers α_1, α_2, α_3, Theorem 1 implies that

$$\alpha_1\pi^{-1/2}\Gamma(\tfrac{1}{4})^2 + \alpha_2\pi^{-1}\Gamma(\tfrac{1}{3})^3 + \alpha_3\pi$$

is either 0 or transcendental. By contrast, it should be pointed out that it is still unknown whether or not $\Gamma(\frac{1}{4})$ or $\Gamma(\frac{1}{3})$ is irrational.

In the following, we only give a detailed proof of Theorem 1. However, it should be reasonably clear to the reader what modifications in the arguments must be made to establish Theorem 2. The essential role of the complex multiplication hypothesis in both proofs is to ensure that the field F_Q appearing in the proof of Lemma 9 has degree $\leqslant cQ^4$ as Q tends to infinity, where c is a number not depending on Q. The validity of this estimate is vital for the transcendence proof in its present form.

2. Preliminary Lemmas.

Throughout this section, $\wp(z)$ will denote a Weierstrass $\wp$-function, whose invariants g_2, g_3 are algebraic numbers. Let ω_1, ω_2 be a basis of the period lattice of $\wp(z)$. As usual, $\mathbb{Q}$ will denote the field of rational numbers, and we let $K = \mathbb{Q}(g_2, g_3)$. Let q be any integer > 1, and let K_q be the field obtained by adjoining to K the four numbers $\wp(\omega_j/q)$, $\wp'(\omega_j/q)(j=1,2)$. In the following, $c_1, c_2, \ldots$ will denote positive numbers which depend only on g_2, g_3, ω_1, ω_2.

LEMMA 1. $e^{\frac{2\pi i}{q}}$ belongs to the field K_q.

For two different proofs, see [5], p. 260, or [7], p. 135.

LEMMA 2. Assume that $\frac{1}{4}g_2, g_3$ are algebraic integers. Then, if λ_1, λ_2 are integers, not both divisible by q, $\wp((\lambda_1\omega_1+\lambda_2\omega_2)/q)$ is an algebraic number with the maximum of the absolute values of its conjugates at most c_1q^2. Further, the leading coefficient in its minimal integral polynomial divides q^{c_2}.
For the proof, we refer the reader to [1].

LEMMA 3. Assume that $\wp(z)$ has complex multiplications. Then the degree of K_q over $\mathbb{Q}$ is at most c_3q^2.

Proof. Let $\tau = \omega_1/\omega_2$. Then τ is an imaginary quadratic irrationality, and so there exists a rational integer b so that $\wp(b\tau z)$ also has periods ω_1, ω_2. Hence

$$\wp(b\tau z) = \Lambda(\wp(z), \wp'(z)) ,$$

where Λ is a rational function of two variables. Further, since g_2, g_3 are algebraic numbers, it follows easily that the coefficients of Λ are algebraic numbers. On putting $z = \omega_2/(bq)$ in this last equation, we conclude that $\wp(\omega_1/q)$ is contained in the field generated over K by $\wp(\omega_2/(bq))$, $\wp'(\omega_2/(bq))$ and the coefficients of Λ. Since $\wp(\omega_2/(bq))$ has degree at most b^2q^2 over K, the assertion of the lemma follows immediately.

Finally, we record three lemmas which will also be used repeatedly in the subsequent arguments. The proofs of these lemmas can be found in either [1] or [2]. We write $\mathcal{S}(z)$ for the

Weierstrass ζ-function associated with $\wp(z)$, and we put $\eta_j = 2\zeta(\omega_j/2)(j=1,2,)$.

<u>LEMMA 4</u>. Assume that $\frac{1}{4}g_2, g_3$ are algebraic integers. Let λ_1, λ_2 be integers with $(\lambda_1,\lambda_2,2q) = 1$. Then, if we put $u = (\lambda_1\omega_1+\lambda_2\omega_2)/(2q)$, the number

$$\xi(\lambda_1,\lambda_2,2q) = (\lambda_1\eta_1 + \lambda_2\eta_2)/(2q) - \zeta(u)$$

belongs to the field generated over K by $\wp(u)$, $\wp'(u)$. Further each of its conjugates has absolute value at most $c_4 q$ and the leading coefficient in its minimal integral polynomial divides $(2q)^{c_5}$.

<u>LEMMA 5.</u> Let λ and m be nonnegative integers. Then

$$\frac{d^m}{dz^m}\wp^\lambda = \sum_{j\geq 0}\ \sum_{j'\geq 0}\ \sum_{\substack{j''\geq 0\\ 2j+3j'+4j''=m+2\lambda}} u(\lambda,m,j)\wp^j\wp'^{j'}\wp''^{j''}$$

where the $u(\lambda,m,j)$ are rational integers with absolute values at most $m!\ c_6^{m+\lambda}$.

<u>LEMMA 6.</u> Let f be a function such that $f'' = -\wp'$. Then, for all nonnegative integers λ and m, we have

$$\frac{d^m}{dz^m} f^\lambda = \sum u(\lambda,m,j,k)f^j f'^{j'}\wp^k\wp'^{k'}\wp''^{k''},$$

where the summation extends over all nonnegative integers j, j', k, k', k'' with

$$j+2j'+2k+3k'+4k'' = m+\lambda \ , \quad j+j' \leq \lambda \ ,$$

and where the $u(\lambda,m,j,k)$ are rational integers with absolute values at most $m!\, c_7^{m+\lambda}$.
In this last lemma, c_7 may, of course, depend on the function f. However, this will not be important for our purposes, since we merely require that c_7 does not depend on m or λ.

3. Proof of Theorem 1.

To simplify the writing of subsequent formulae, we now use a different notation from that given in §1. We denote our two Weierstrass $\wp$-functions by $\wp_1(z)$, $\wp_3(z)$, their corresponding ζ-functions by $\zeta_1(z)$, $\zeta_3(z)$, and their invariants by $g_{2,\ell}$, $g_{3,\ell}$ $(\ell=1,3)$. We let ω_1, ω_2 be a basis of the period lattice of $\wp_1(z)$, and ω_3, ω_4 a basis of the period lattice of $\wp_3(z)$. Put $\eta_j = 2\,\zeta_1(\omega_j/2)(j=1,2)$, and $\eta_j = 2\zeta_3(\omega_j/2)(j = 3,4)$.

We assume that both $\wp_1(z)$ and $\wp_3(z)$ have complex multiplications. We also suppose, of course that the $g_{2,\ell}$, $g_{3,\ell}(\ell=1,3)$ are algebraic numbers. The proof of Theorem 1 will be by contradiction. Thus we suppose that there exist algebraic integers $\alpha_o \neq 0$, α_j, β_j $(1 \leq j \leq 4)$, α_5 such that

$$(3) \qquad \alpha_o = \sum_{j=1}^{4} (\alpha_j\omega_j + \beta_j\eta_j) + \alpha_5 2\Pi i \ ,$$

and we shall ultimately show that this is impossible. We first remark that there is no loss of generality in assuming that $\frac{1}{4}g_{2,\ell}$, $\frac{1}{4}g_{3,\ell}$ $(\ell=1,3)$ are algebraic integers. For, if this is not the case, we simply work with the Weierstrass functions with invariants $a^4 g_{2,\ell}$, $a^6 g_{3,\ell}$ $(\ell=1,3)$ for an appropriate rational integer a.

The following additional notation will be used throughout the proof. We also use $\wp_2(z)$, $\wp_4(z)$ to denote $\wp_1(z)$, $\wp_3(z)$, respectively, and similarly for the associated ζ-functions. Let F be the field generated over $\mathbb{Q}$ by

$$(4) \quad \alpha_j (0 \leqslant j \leqslant 5), \quad \beta_j (1 \leqslant j \leqslant 4), \quad g_{2,\ell}, \; g_{3,\ell}, \; (\ell=1,3),$$

and the numbers $\wp_j^{(\ell)}(\omega_j/2)$ $(1 \leqslant j \leqslant 4,\ 0 \leqslant \ell \leqslant 1)$. Let $d = [F:\mathbb{Q}]$. If λ, μ are nonnegative integers, we write $\wp_j(\lambda,\mu,z)$ for the μ-th derivative of the λ-th power of $\wp_j(z)$. By $c_1, c_2, \ldots$ we signify positive numbers which depend only on the numbers (4) and the ω_j $(1 \leqslant j \leqslant 4)$. A basic role in the proof will be played by a function of 5 complex variables $z_1, \ldots, z_5$, and, if $f(z_1, \ldots, z_5)$ is a meromorphic function of these variables, we put

$$f_{m_1,\ldots,m_5}(z_1,\ldots,z_5) = \frac{\partial^{m_1+\ldots m_5}}{\partial z_1^{m_1} \ldots \partial z_5^{m_5}} f(z_1,\ldots,z_5)$$

for all nonnegative integers $m_1, \ldots, m_5$.

The basic parameter in our subsequent arguments will be denoted by k, and we put

$$h = [k^{1/14}], \qquad L = [k^{6/7}].$$

Throughout the proof it will be assumed that k is sufficiently large. Finally, we let

$$f(z_1,\ldots,z_5) = \sum_{j=1}^{4} (\alpha_j\omega_j z_j + \beta_j \wp_j(\omega_j z_j)) + \alpha_5 2\pi i\, z_5 .$$

For simplicity, we shall often write just f instead of $f(z_1,\ldots,z_5)$.

<u>LEMMA 7.</u> There exist rational integers $p(\lambda_0,\ldots,\lambda_5)$, not all 0, with absolute values at most k^{2k}, such that the function

$$\Phi(z_1,\ldots,z_5) = \sum_{\lambda_0=0}^{L} \cdots \sum_{\lambda_5=0}^{L} p(\lambda_0,\ldots,\lambda_5) f^{\lambda_0} \left(\prod_{j=1}^{4} \wp_j(\omega_j z_j)^{\lambda_j}\right) e^{2\pi i \lambda_5 z_5}$$

satisfies

$$\Phi_{m_1,\ldots,m_5}(s+\tfrac{1}{2},\ldots,s+\tfrac{1}{2}) = 0 \tag{5}$$

for all integers s with $1 \leqslant s \leqslant h$, and all non-negative integers $m_1,\ldots,m_5$ with $m_1+\ldots+m_5 \leqslant k$.

<u>Proof.</u> The expression on the left of (5) is given by

$$\sum_{\lambda_0=0}^{L} \cdots \sum_{\lambda_5=0}^{L} p(\lambda_0,\ldots,\lambda_5) \left\{ \sum_{\mu_1=0}^{m_1} \cdots \sum_{\mu_5=0}^{m_5} \binom{m_1}{\mu_1} \cdots \binom{m_5}{\mu_5} \Lambda \prod_{j=1}^{5} P_j \right\}, \tag{6}$$

where Λ denotes $\dfrac{\partial^{\mu_1+\dots+\mu_5}}{\partial z_1^{\mu_1}\dots\partial z_5^{\mu_5}} f^{\lambda_o}$ evaluated at $z_1 = \dots = z_5 = s+\frac{1}{2}$, and where

$$P_j = \omega_j^{m_j} \wp_j(\lambda_j, m_j-\mu_j, \tfrac{1}{2}\omega_j) \, (1 \leq j \leq 4) ,$$
$$P_5 = (2\pi i\lambda_5)^{m_5-\mu_5}(-1)^{\lambda_5} .$$

Put $\wp_{\ell j} = \wp_j^{(\ell)}(\omega_j/2)(1=0,1,2)$. We see from Lemma 5 that

$$\omega_j^{-(m_j-\mu_j)} P_j = \sum_j u_j \wp_{oj}^{t_{oj}} \wp_{1j}^{t_{1j}} \wp_{2j}^{t_{2j}} \qquad (1 \leq j \leq 4) ,$$

where $\sum_j$ denotes summation over all nonnegative integers t_{oj}, t_{1j}, t_{2j} with

$$2t_{oj} + 3t_{1j} + 4t_{2j} = m_j - \mu_j + 2\lambda_j ,$$

and $u_j = u_j(t_{oj}, t_{1j}, t_{2j}, m_j-\mu_j, \lambda_j)$ denotes a rational integer with absolute value at most $(m_j-\mu_j)! \, c_1^{m_j+2\lambda_j}$. Now it is easily seen that our hypothesis that the $\frac{1}{4}g_{2,\ell}$, $\frac{1}{4}g_{3,\ell}$ are algebraic integers implies that all of the $\wp_{\ell j}$ are algebraic integers. Hence, since $m_j \leq k$ and $L < k$, we conclude that $\omega_j^{-(m_j-\mu_j)} P_j$ $(1 \leq j \leq 4)$ is an algebraic integer in F with the maximum of the absolute values of its conjugates at most

$$(m_j + 2L)^3 c_2^{m_j+2L} (m_j - \mu_j)! \leq (m_j - \mu_j)! c_3^k .$$

Also, as $\lambda_5^{m_5-\mu_5} \leq 2^{m_5+L}(m_5-\mu_5)!$, we see that

$(2\pi i)^{-(m_5-\mu_5)}P_5$ is a rational integer with absolute value at most $(m_5-\mu_5)!c_4^k$. We next consider Λ. For brevity, let us write $\wp_j$ for $\wp_j(\omega_j z_j)$ $(1\leq j\leq 4)$. Then, by virtue of Lemma 6, the μ_1-th derivative of f^{λ_o} with respect to z_1 can be written in the form

$$(7)\qquad \omega_1^{\mu_1}\sum_1{}^* u_1^* v_1^* \beta_1^{\lambda_o-\tau_1-\tau_1'}f^{\tau_1} ,$$

where

$$v_1^* = (\alpha_1 - \beta_1\wp_1)^{\tau_1'}\wp_1^{t_1}\wp_1'^{t_1'}\wp_1''^{t_1''} ,$$

the summation is over all nonnegative integers $\tau_1,\tau_1',t_1,t_1',t_1''$ with

$$\tau_1 + 2\tau_1' + 2t_1 + 3t_1' + 4t_1'' = \mu_1 + \lambda_o ,$$

$$\tau_1 + \tau_1' \leq \lambda_o ,$$

and $u_1^* = u_1^*(\tau_1,\tau_1',t_1,t_1',t_1'',\mu_1,\lambda_o)$ denotes a rational integer with absolute value at most $\mu_1!\,c_5^{\mu_1+L}$. For $j = 2,\ldots,4$, let $\sum_j^*$ denote summation over all nonnegative integers τ_j, τ_j', t_j, t_j', t_j'' satisfying

$$\tau_j + 2\tau_j' + 2t_j' + 3t_j' + 4t_j'' = \mu_j + \tau_{j-1} ,$$

$$\tau_j + \tau_j' \leq \tau_{j-1} .$$

Then, applying Lemma 6 again, we conclude that the μ_2-th derivative of (7) with respect to z_2 can be written in the form

$$(8)\qquad \begin{aligned} &\omega_1^{\mu_1}\omega_2^{\mu_2} \sum_1{}^{*} \sum_2{}^{*} u_1^* u_2^* v_1^* v_2^* \\ &\beta_1^{\lambda_o-\tau_1-\tau_1'}\beta^{\tau_1-\tau_2-\tau_2'} f^{\tau_2}, \end{aligned}$$

where v_2^* is defined by v_1^* above with 1 replaced by 2 throughout, and $u_2 = u_2^*(\tau_2,\tau_2',t_2,t_2',t_2'',\mu_2,\tau_1)$ denotes a rational integer with absolute value at most $\mu_2!\, c_5^{\mu_2+L}$. For simplicity, put $\tau_o = \lambda_o$. Applying Lemma 6 once more, we see that the μ_3-th derivative of (8) with respect to z_3 is

$$(9)\quad \omega_1^{\mu_1}\omega_2^{\mu_2}\omega_3^{\mu_3} \sum_1{}^{*} \sum_2{}^{*} \sum_3{}^{*} \prod_{\ell=1}^{3} (u_\ell^* v_\ell^* \beta_\ell^{\tau_{\ell-1}-\tau_\ell-\tau_\ell'}) f^{\tau_3},$$

where v_3^* is defined by v_1^* above with 1 replaced by 3 throughout, and $u_3^* = u_3^*(\tau_3,\tau_3',t_3,t_3',t_3'',\mu_3,\tau_2)$ denotes a rational integer with absolute value at most $\mu_3!\, c_5^{\mu_3+L}$. Repeating this again, the μ_4-th derivative of (9) with respect to z_4 is

$$(10)\quad \omega_1^{\mu_1}\cdots\omega_4^{\mu_4} \sum_1{}^{*}\cdots\sum_4{}^{*} \prod_{\ell=1}^{4} (u_\ell^* v_\ell^* \beta_\ell^{\tau_{\ell-1}-\tau_\ell-\tau_\ell'}) f^{\tau_4},$$

where v_4^* is defined by v_1^* above with 1 replaced by 4 throughout, and $u_4^* = u_4^*(\tau_4,\tau_4',t_4,t_4',t_4'',\mu_4,\tau_3)$ denotes a rational integer with absolute value at most $\mu_4!\, c_5^{\mu_4+L}$. Finally, the μ_5-th derivative of (10) with respect to z_5 is

$$\omega_1^{\mu_1}\ldots\omega_4^{\mu_4}(2\pi i)^{\mu_5}\ \sum_1{}^*\ldots\sum_4{}^*\ \prod_{\ell=1}^{4}\left(u_\ell^* v_\ell^* \beta_\ell^{\tau_{\ell-1}-\tau_\ell-\tau'_\ell}\right)$$

$$\frac{\tau_4!}{(\tau_4-\mu_5)!} f^{\tau_4-\mu_5} .$$

Now, by virtue of the equation $\zeta_j(z+\omega_j) = \zeta_j(z) + \eta_j$, we have in particular that $\zeta_j((s+\frac{1}{2})\omega_j) = (s + \frac{1}{2})\eta_j$. Hence we conclude from (3) that $f(s+\frac{1}{2},\ldots,s+\frac{1}{2}) = (s+\frac{1}{2})\alpha_o$, whence

$$\Lambda = \omega_1^{\mu_1}\ldots\omega_4^{\mu_4}(2\pi i)^{\mu_5}\ \sum_1{}^*\ldots\sum_4{}^*\ \prod_{\ell=1}^{4}$$

(11)

$$\left(u_\ell^* w_\ell^* \beta_\ell^{\tau_{\ell-1}-\tau_\ell-\tau'_\ell}\right)\ \frac{\tau_4!}{(\tau_4-\mu_5)!}\ (\alpha_o(s+\tfrac{1}{2}))^{\tau_4-\mu_5} ,$$

where, recalling that $\wp_{gj} = \wp_j^{(g)}(\omega_j/2)$,

$$w_\ell^* = (\alpha_\ell - \beta_\ell\,\wp_{o\ell})^{\tau'_\ell}\,\wp_{o\ell}^{t_\ell}\,\wp_{1\ell}^{t'_\ell}\,\wp_{2\ell}^{t''_\ell}\quad(\ell=1,\ldots,4).$$

Thus, as the α_ℓ, β_ℓ , and $\wp_{g\ell}$ are algebraic integers, it is plain that

$$\omega_1^{-\mu_1}\ldots\omega_4^{-\mu_4}(2\pi i)^{-\mu_5}\ 2^L\ \Lambda$$

is an algebraic integer in F with the maximum of the absolute values of its conjugates at most

$$(k+L)^{20} s^L c_6^{k+L}\mu_1!\ \ldots\ \mu_5! \leqslant h^L c_7^k \mu_1!\ \ldots\ \mu_5!\ .$$

Combining this last result with our previous estimates for the P_j $(1 \leqslant j \leqslant 5)$, and recalling that

$$(m_j-\mu_j)!\ \mu_j! \leqslant m_j!\ ,\ \binom{m_j}{\mu_j} \leqslant 2^{m_j}\ ,$$

$$\prod_{j=1}^{5}(m_j!) \leqslant (\sum_{j=1}^{5} m_j)! \leqslant k^k\ ,$$

we conclude that

$$\Omega(m,\lambda) = \omega_1^{-m_1}\cdots\omega_4^{-m_4}(2\pi i)^{-m_5}\sum_{\mu_1=0}^{m_1}\cdots\sum_{\mu_5=0}^{m_5}$$

$$\binom{m_1}{\mu_1}\cdots\binom{m_5}{\mu_5} 2^L \Lambda \prod_{j=1}^{5} P_j$$

is an algebraic integer in F with the maximum of the absolute values of its conjugates at most $k^k h^L c_8^k$. Hence, denoting by $\Theta_1,\ldots,\Theta_n$ an integral basis of F over $\mathbb{Q}$, we have

$$\Omega(m,\lambda) = \sum_{\ell=1}^{d} \nu_\ell(m,\lambda)\Theta_\ell\ ,$$

where the $\nu_\ell(m,\lambda)$ are rational integers with absolute values at most $c_q^k k^k h^L$. It follows that (5) will be satisfied if the d equations

$$\sum_{\lambda_0=0}^{L}\cdots\sum_{\lambda_5=0}^{L} p(\lambda_0,\ldots,\lambda_5)\nu_\ell(m,\lambda) = 0 \quad (1\leqslant l\leqslant d)$$

hold for all s with $1 \leqslant s \leqslant h$, and all nonnegative integers $m_1,\ldots,m_5$ with $m_1+\ldots+m_5 \leqslant k$. This gives $M \leqslant dh(k+1)^5$ equations in the $N = (L+1)^6$ unknowns $p(\lambda_0,\ldots,\lambda_5)$, and, since $h \leqslant k^{1/14}$ and $L+1 > k^{6/7}$, we see that $N > k^{36/7} > 2M$. Thus, by Siegel's lemma on linear equations (cf.

Lemma 8 of [1]), these equations have a nontrivial solution in integers $p(\lambda_0,\ldots,\lambda_5)$ with absolute values at most NU, where $U = c_q^{5k} k^k h^L$. On noting that $NU \leqslant k^{2k}$ because $L \leqslant k^{6/7}$, the proof of Lemma 7 is complete.

Now suppose that R is a positive real number, which will eventually be taken to be very large. For simplicity, we assume in the following that $R \geqslant 6$. We define

$$\Psi(z_1,\ldots,z_5) = \Phi(z_1,\ldots,z_5) \prod_{j=1}^{4} \prod_{\Omega_j} (\omega_j z_j - \Omega_j)^{3L},$$

where Ω_j runs over all periods of $\wp_j(z)$ with $|\Omega_j| \leqslant |\omega_j| R$. The purpose in taking this definition is, of course, to insure that $\Psi(z_1,\ldots,z_5)$ is regular in the disc $|z_j| \leqslant R$ $(1 \leqslant j \leqslant 5)$.

<u>LEMMA 8</u>. For any z with $|z| \leqslant R/2$, and any nonnegative integers $m_1,\ldots,m_5$ with $m_1+\ldots+m_5 \leqslant k$, we have

$$|\Psi_{m_1,\ldots m_5}(z,\ldots,z)| \leqslant k^{4k} R^{c_{10} LR^2}.$$

<u>Proof</u>. For $j = 1,\ldots,4$, the number of Ω_j occurring in the product defining $\Psi(z_1,\ldots,z_5)$ is at most $c_{11}R^2$. Hence, since $\wp_j(z)$ and $\zeta_j(z)$ have double and single poles, respectively, at the points of their period lattice, we have

$$|\wp_j(\omega_j z_j) \prod_{\Omega_j} (\omega_j z_j - \Omega_j)^2| \leqslant R^{c_{12}R^2},$$

$$|\zeta_j(\omega_j z_j) \prod_{\Omega_j} (\omega_j z_j - \Omega_j)| < R^{c_{12}R^2}$$

for $|z_j| \leq 2R/3$, say. Thus

(12) $$|\Psi(z_1,\ldots,z_5)| \leq (L+1)^6\, k^{2k} R^{c_{13}LR^2},$$

in the disc $|z_j| \leq 2R/3$ $(1 \leq j \leq 5)$. Now, if $|z| \leq R/2$ and if $m_1,\ldots,m_5$ are nonnegative integers with $m_1+\ldots+m_5 \leq k$, we have

$$\Psi_{m_1,\ldots,m_5}(z,\ldots,z) = \frac{m_1!\ldots m_5!}{(2\pi i)^5} \int_{C_1} \cdots \int_{C_5} \frac{\Psi(z_1,\ldots,z_5)\, dz_1\ldots dz_5}{\prod_{j=1}^{5} (z_j - z)^{m_j+1}},$$

where C_j denotes the positively oriented circle in the complex plane with centre the origin and radius $2R/3$. But $|z_j - z| \geq R/6 \geq 1$ for z_j on C_j, and also $\prod_{j=1}^{5}(m_j!) \leq k^k$, whence the estimate asserted follows at once from (12).

<u>LEMMA 9.</u> Let Q,S be numbers satisfying $1 < Q < S < R-1$. Let $m_1,\ldots,m_5$ be nonnegative integers satisfying $m_1+\ldots+m_5 \leq k$. Suppose that q,r,s are integers, with q even, $(r,q) = 1$, and

$$1 \leq q \leq Q, \quad 1 \leq s \leq S, \quad 1 \leq r < R$$

such that

(13) $$\Psi_{\mu_1,\ldots,\mu_5}(s+r/q,\ldots,s+r/q) = 0$$

for all nonnegative integers $\mu_1,\ldots,\mu_5$ with $\sum_{j=1}^{5} \mu_j \leqslant \sum_{j=1}^{5} m_j$. Then either (13) holds when $\mu_j = m_j$ $(1 \leqslant j \leqslant 5)$, or we have

$$(14)\quad |\Psi_{m_1,\ldots,m_5}(s+r/q,\ldots,s+r/q)| > (kS)^{-c_{14}kQ^4} .$$

Proof. The hypotheses of the lemma imply that when $\mu_j = m_j$ $(1 \leqslant j \leqslant 5)$, the number on the left of (14) is given by $W\Psi$, where

$$W = \prod_{j=1}^{5} \prod_{\Omega_j} (\omega_j(s+r/q) - \Omega_j)^{3L} ,$$

$$\Psi = \Psi_{m_1,\ldots,m_5}(s+r/q,\ldots,s+r/q) .$$

We claim that $|W| \geqslant (c_{15}/Q)^{c_{16}L}$. For the absolute value of $\omega_j(s+r/q) - \Omega_j$ clearly exceeds c_{17} unless $\Omega_j = s\omega_j$ or $(s+1)\omega_j$; in the latter two cases we have the lower bound c_{18}/Q. Thus, to prove the lemma, it suffices to show that $|\Psi| > (kS)^{-c_{20}kQ^4}$.

We now define

$$(15)\quad \wp_{\ell j} = \wp_j^{(\ell)}(r\,\omega_j/q) \qquad (0 \leqslant l \leqslant 2,\quad 1 \leqslant j \leqslant 4) .$$

It is plain from (3) that

$$(16)\quad f(s+r/q,\ldots,s+r/q) = \alpha_0(s+r/q) - \sum_{j=1}^{4} \beta_j \xi_j ,$$

where $\xi_j = r\,\eta_j/q - \zeta_j(r\,\omega_j/q)$. Now clearly Ψ is given by (6), but with the following new definitions of $\Lambda, P_1,\ldots,P_5$. In the expression (11) for Λ, replace $\alpha_o(s+\frac{1}{2})$ by (16), and take the new value (15) of $\wp_{\ell j}$. In the definition of P_j $(1 \leqslant j \leqslant 4)$, replace $\omega_j/2$ by $r\,\omega_j/q$. Finally, in the definition of P_5, replace -1 by $e^{\frac{2\pi i r}{q}}$. Hence, if F_q denotes the field generated over $\mathbb{Q}$ by the numbers (4) and the numbers $e^{\frac{2\pi i}{q}}$, $\wp_j(\ell)(\omega_j/q)(0 \leqslant \ell \leqslant 1,\ 1 \leqslant j \leqslant 4)$, we conclude from Lemma 4 that

$$\Psi' = \omega_1^{-m_1} \ldots \omega^{-m_4}(2\pi i)^{-m_5}\Psi$$

is an algebraic number lying in F_q. By Lemmas 1 and 3, F_q has degree at most $c_{21}Q^4$ over the rational field $\mathbb{Q}$. Further, it follows easily from Lemmas 2 and 4 that we can find an integer c_{22} such that $q^{c_{22}}$ times the numbers (15) and (16) are both algebraic integers with the maximum of the absolute values of their conjugates at most $q^{c_{23}}$. We then conclude from the explicit expression for Ψ' derived in the proof of Lemma 7 that

$$q^{c_{24}(m_1+\ldots+m_5+L)}\Psi' \tag{17}$$

is an algebraic integer with the maximum of the absolute values of its conjugates at most

$$(L+1)^6 k^{2k} 2^k m_1! \ldots m_5!(c_{24}S)^{c_{25}(k+L)} < (kS)^{c_{26}k}.$$

Since Ψ' has degree at most $c_{21}Q^4$, the assertion of the lemma follows on noting that either $\Psi' = 0$ or the norm of (17) is at least 1.

The reader may find the following comment useful in connection with the next lemma. The value $J = 50$ given in the lemma is chosen simply because it is sufficiently large to ensure that $h^{J/4+1}$ exceeds $L+1$ (this latter estimate is plainly vital for the determinant argument occurring at the end of the paper). In fact, the assertion of the lemma remains valid for any choice of J which does not depend on k.

LEMMA 10. Let J be any integer satisfying $0 \leqslant J \leqslant 50$. Then

$$\Phi_{m_1,\ldots,m_5}(s+r/q,\ldots,s+r/q) = 0$$

for all integers q,r,s, with q even, $(r,q)=1$,

$$1 \leqslant q \leqslant 2h^{J/8}, \quad 1 \leqslant s \leqslant h^{J/4+1}, \quad 1 \leqslant r < q ,$$

and all nonnegative integers $m_1,\ldots,m_5$ with $m_1+\ldots+m_5 \leqslant k/2^J$.

Proof. The lemma is valid for $J = 0$ by Lemma 7. We suppose that I is an integer with $0 \leqslant I < 50$, and we assume that the lemma holds for $J = 0,\ldots,I$. We now proceed to deduce the assertion of the lemma for $J = I + 1$.

The proof is by contradiction. Thus if we put

$$Q_j = 2h^{J/8}, \quad S_J = h^{J/4+1}, \quad T_J = [k/2^J]$$

$$(J = 0,1,\ldots) ,$$

we assume that there exist integers q', r', s',

with q' even, $(r',q') = 1$,

$$1 \leq q' \leq Q_{I+1}, \quad 1 \leq s' \leq S_{I+1}, \quad 1 \leq r' < q' ,$$

and nonnegative integers $m_1',\ldots,m_5'$ with $m'_1+\ldots+m_5' \leq T_{I+1}$, satisfying

$$\Phi_{m_1',\ldots,m_5'}(s'+r'/q',\ldots,s'+r'/q') \neq 0 .$$

We can also plainly suppose that $m_1',\ldots,m_5'$ are chosen minimally in the sense that

$$\Phi_{\mu_1,\ldots,\mu_5}(s'+r'/q',\ldots,s'+r'/q') = 0$$

for all nonnegative integers $\mu_1,\ldots,\mu_5$ with $\sum_{j=1}^{5} \mu_j < \sum_{j=1}^{5} m_j'$.

Now take the number R introduced before Lemma 8 to be $10\, S_{I+1}$, and let $\Psi(z) = \Psi_{m_1',\ldots,m_5'}(z,\ldots,z)$. Then it follows easily from our inductive hypothesis (cf. proof of Lemma 10 in [3]) that for all integers q,r,s, with q even, $(r,q) = 1$,

$$1 \leq q \leq Q_I, \quad 1 \leq s \leq S_I, \quad 1 \leq r < q ,$$

and each integer m satisfying $0 \leq m \leq T_{I+1}$, we have

$$\frac{d^m}{dz^m}\Psi(s+r/q) = 0. \tag{18}$$

Hence, if we write

$$F(z) = \prod_{\substack{q=2\\ q \text{ even}}}^{Q_I} \prod_{s=1}^{S_I} \prod_{\substack{t=1\\ (T,q)=1}}^{q} (z-s-r/q)^{T_{I+1}},$$

the function $\Psi(z)/F(z)$ is certainly regular in the disc $|z| \leqslant 5S_{I+1}$. Also, $s'+r'/q'$ plainly lies in this disc. Hence, denoting by θ and Θ the upper bound of $|\Psi(z)|$ and the lower bound of $|F(z)|$, respectively, on the circle $|z| = 5S_{I+1}$, we conclude from the maximum modulus principle that

$$(19) \qquad |\Psi(s'+r'/q')| \leqslant |F(s'+r'/q')|\theta/\Theta .$$

Now, by Lemma 8, we have

$$(20) \qquad \theta \leqslant k^{4k}(10S_{I+1})^{c_{10}LS_{I+1}^2} .$$

Further, since for any z with $|z| = 5S_{I+1}$,

$$|z-s-r/q| \geqslant 2|(s'+r'/q') - (s+r/q)| ,$$

and since the number of triplets q,r,s which occur in the definition of $F(z)$ is at least $c_{27}Q_I^2S_I$ (cf. proof of Lemma 10 of [3]), we clearly have

$$(21) \qquad \Theta > 2^{c_{28}Q_I^2S_IT_{I+1}} |F(s'+r'/q')| .$$

It now follows readily from (19), (20) and (21) that

$$(22) \qquad |\Psi(s'+r'/q')| < 2^{-c_{29}Q_I^2S_IT_{I+1}} .$$

On the other hand, we deduce from Lemma 9 and our initial hypothesis that $\Psi(s'+r'/q') \neq 0$, that

$$(23) \qquad |\Psi(s'+r'/q')| > (kS_{I+1})^{-c_{30}kQ_{I+1}^4} .$$

But the estimates (22) and (23) are contradictory

when k is sufficiently large (cf. proof of Lemma 10 of [3]). This contradiction completes the proof of Lemma 10.

We can now complete the proof of Theorem 1. From Lemma 10, we have

$$(24) \qquad \Phi_{m_1,\ldots,m_5}(s+\tfrac{1}{4},\ldots,s+\tfrac{1}{4}) = 0$$

for all integers s with $1 \leq s \leq L+1$, and all integers $m_1,\ldots,m_5$ with $0 \leq m_1,\ldots,m_5 \leq L$. By an entirely similar argument to that given at the end of [3], we conclude from (24) that

$$(25) \quad q(\lambda_o,\nu_1,\ldots,\nu_5) = 0 \quad (0 \leq \lambda_o \leq L,\ 0 \leq \nu_1,\ldots,\nu_5 \leq L)$$

where

$$q(\lambda_o,\nu_1,\ldots,\nu_5) = \sum_{\lambda_1=0}^{L} \cdots \sum_{\lambda_5=0}^{L} p(\lambda_o,\ldots,\lambda_5)$$

$$\mathcal{G}_1(\lambda_1,\nu_1,\tfrac{1}{4}\,\omega_1)\ldots \times \mathcal{G}_4(\lambda_4,\nu_4,\tfrac{1}{4}\,\omega_4)\lambda_5^{\nu_5}(i)^{\lambda_5} .$$

Now choose λ_o so that $p(\lambda_o,\ldots,\lambda_5) \neq 0$ for some $\lambda_1,\ldots,\lambda_5$. It follows from (25) that the determinant Δ of the $(L+1)^5$ equations

$$q(\lambda_o,\nu_1,\ldots,\nu_5) = 0 \qquad (0 \leq \nu_1,\ldots,\nu_5 \leq L)$$

must vanish. On the other hand, the determinant Δ is given by

$$\Delta = (\Delta_1 \ldots \Delta_5)^{L+1}$$

where

$$\Delta_j = 2!\ldots L!(\wp_j'(\tfrac{1}{4}\omega_j))^{\frac{1}{2}L(L+1)} \quad (j=1,\ldots,4)$$

$$\Delta_5 = 2!\ldots L!(-2\pi)^{\frac{1}{2}L(L+1)}$$

(cf. Lemmas 6 and 7 of [1]). But $\wp_j'(\frac{1}{4}\omega_j) \neq 0$, and thus $\Delta \neq 0$. This contradiction shows that (3) cannot hold, and so the proof of Theorem 1 is complete.

REFERENCES

[1] A. Baker, On the periods of the Weierstrass $\wp$-function, Symposia Mathematica, Instituto Nazionale Alta Matematica, Vol. IV., Rome (1970).

[2] A. Baker, On the quasi-periods of the Weierstrass $\wp$-function, Nach. Akad. Wiss. Gottingen, Math. Phys. Klasse, Nr. 16 (1969).

[3] J. Coates, The transcendence of linear forms in $\omega_1, \omega_2, \eta_1, \eta_2, 2\pi i$, American Journal Math., 113 (1971), 385-397.

[4] J. Coates, Linear forms in the periods of the exponential and elliptic functions, Inventiones Math., 12 (1971), 290-299.

[5] R. Fricke, Die elliptischen Funktionen und ihre Anwendungen, Teil 2, (Leipzig and Berlin, 1922).

[6] T. Schneider, Einfuhrung in die transzendenten Zahlen (Berlin, 1957).

[7] G. Shimura, Introduction to the arithmetic theory of automorphic functions, Publications of the Math. Soc. of Japan, Nr. 11, (1971).

Harvard University, Cambridge, Massachusetts

This research was supported in part by the United States Army Office of Research (Durham).

THE LARGEST COEFFICIENT IN CONTINUED FRACTIONS AND RELATED PROBLEMS

Janos Galambos

1. Introduction

The coefficients $a_1(x), a_2(x), \ldots$ in the regular continued fraction expansion of $x \in (0,1)$ are defined as follows. Let $Tx = 1/x \pmod 1$; then

$$a_1(x) = 1/x - Tx, \quad a_{j+1}(x) = a_j(Tx) = a_1(T^j x) .$$

Let

$$L_N = L_N(x) = \max(a_1(x), a_2(x), \ldots, a_N(x)) .$$

Let us denote by P Gauss measure defined by

$$P(A) = \frac{1}{\log 2} \int_A \frac{dx}{1+x}$$

for all Lebesgue measurable sets A and let $\lambda(A)$ be the Lebesgue measure of A. In my paper Galambos (1972a), the P-distribution of L_N is investigated and I found the suitable normalizing constant for a limit law. In this paper we extend the result just mentioned to any measure absolutely continuous with respect to P which include two noteworthy special cases (i) Lebesgue measure (ii) Lebesgue measure or P-measure restricted to a given subset of the interval (0,1). From this we shall deduce that L_N, in whatever way normalized, cannot have positive limit on a set of positive Lebesgue measure. Finally, some remarks will be made on the metric properties of the remainders of a large class of number theoretical transformations, in the light of the present investigation.

As pointed out in Galambos (1972a), the P-distribution of L_N could have been investigated through the general theory of extreme value distributions of Galambos (1972b). The concept of stochastic dependence employed there would, however, have made our studies unnecessarily complicated.

2. <u>The Distribution of L_N</u>. Our results on L_N are formulated in the following two theorems.

<u>THEOREM 1</u>. Let Q be absolutely continuous with respect to Gauss measure P. Then

$$\lim_{N=+\infty} Q\left(\frac{L_N}{N} < \frac{y}{\log 2}\right) = \exp\left(-\frac{1}{y}\right)$$

<u>THEOREM 2</u>. Let G_N be a sequence of real numbers such that L_N/G_N has a finite limit, as $N \to +\infty$, which limit is positive on a set A. Then $\lambda(A) = 0$.

For Q = P, Theorem 1 was obtained in Galambos (1972a). As pointed out in the introduction, the present extension to an arbitrary measure Q which is absolutely continuous with respect to P implies that the same limit relation holds on any subset of $(0,1)$ and also that P can be replaced by Lebesgue measure. Theorem 2 follows from Theorem 1 but it is not well known even among probabilists. It implies that the weak convergence of Theorem 1 cannot be extended to a pointwise (almost everywhere) convergence.

Proof of Theorem 1. Denote by C_k the set $\{L_k < kz\}$, where z is a fixed real number (not depending on x). By the main result of Rényi (1963), the conclusion of Theorem 1 follows if we show that, as $N \to +\infty$,

$$(1) \quad \lim P(C_N | C_k) = \lim P(C_N) = \exp(-\frac{1}{y}) ,$$

for all fixed k, where $z \log 2 = y$ and $P(C|D)$ denotes conditional probabilities. The last equation in (1) is the result of Galambos (1972a), hence we have only to show the equation with the outer terms in (1). This equation can be written into the equivalent form

$$(2) \lim P(C_N \; C_k) = \exp(-\frac{1}{y})P(C_k), \quad (N \to +\infty) ,$$

with k fixed and $z \log 2 = y$. For evaluating the limit on the left hand side of (2) we shall employ a sieve argument and make use of the mixing property of the coefficients $a_j(x)$. Put $A_{j,N} = \{a_j(x) \geq Nz\}$ and let $B_{k,N}$ be the intersection of the complements of $A_{j,N}$, $k+1 \leq j \leq N$. Then the well known Poincaré identity yields that

$$(3) \qquad P(C_N\ C_k) = P(C_k) + \sum_{t=1}^{N-k} (-1)^t S_t$$

where

$$S_t = \sum P(C_k\ A_{j_1,N}\ A_{j_2,N} \cdots A_{j_t,N}) \ ,$$

the summation being extended over all t - tuples $(j_1, j_2, \ldots, j_t)$ with $k = j_0 < j_1 < j_2 < \ldots < j_t \leq N$. The terms S_t can easily be evaluated by making use of the well known facts that the P-distribution of $a_j(x)$ does not depend on j and that the a's form a mixing sequence. More precisely, it is known (see Philipp (1970)) that

$$P(A_{j,N}) = \frac{\log(1 + 1/[Nz])}{\log 2} = p_N \ , \quad \text{say},$$

and that for any sets $E \in M_{1,s}$ and $F \in M_{s+n,+\infty}$

$$(4) \qquad |P(EF) - P(E)P(F)| < cq^n\ P(E)P(F) \ ,$$

where $0 < q < 1$ and $c > 0$ is a constant and where $M_{u,v}$ is the smallest σ-algebra with respect to which the coefficients $a_u(x)$, $a_{u+1}(x)$, $\ldots, a_v(x)$ are measurable. Since from this point on, the calculations required to complete the proof are similar to those in Galambos (1972a), I omit some minor details. Let us first observe that, by the repeated application of (4) we have, for any choice of t different subscripts,

$$(5)\qquad \begin{aligned} &P(C_k\, A_{j_1,N}\, A_{j_2,N}\cdots A_{j_t,N}) \\ &\leqslant (1+c)^t P(C_k)P(A_{j_1,N})\cdots P(A_{j_t,N}) \\ &\leqslant (1+c)^t p_N^t \,. \end{aligned}$$

On the other hand, we can get asymptotic expression for the left hand side of (5), when the subscripts satisfy the restriction $j_{s+1}-j_s \geqslant n$, $0 \leqslant s \leqslant t-1$, where n is determined so that $cq^n < \varepsilon$. For these choices we have from (4) and (5), that

$$(6)\qquad P(C_k\, A_{j_1,N}\cdots A_{j_t,N}) = P(C_k)p_N^t(1+O(\varepsilon)) \;,$$

the constant in $O(\varepsilon)$ depending on t only. Splitting now the terms of S_t into two classes according as (6) is applicable or not, we get from (5) and (6) that, for any fixed t,

$$(7)\qquad S_t = (1+O(\varepsilon))\,\frac{1}{y^t\, t!}\,P(C_k) + o_N(1) \;,$$

where the last term tends to 0 as $N \to +\infty$. (3) and (7) thus yield

$$(8)\qquad \begin{aligned} P(C_N\, C_k) &= P(C_k)\sum_{t=0}^{Z}(-1)^t(1+O(\varepsilon))\,\frac{1}{y^t\, t!} \\ &\quad + o_N(1) + O\Big(\sum_{t>Z} S_t\Big) \;, \end{aligned}$$

with arbitrary fixed Z (not depending on N). In view of (5), however, $\sum S_t$, t running from Z, can be made arbitrarily small for Z large. Thus letting N tend to $+\infty$, and then Z to $+\infty$, (8) reduces to (2), and Theorem 1 is thus established.

Proof of Theorem 2. From Theorem 1 it follows that if $G_N = o(N)$, then $L_N/G_N \to +\infty$ for almost all x. On the other hand, if $N = o(G_N)$, then $L_N/G_N \to 0$ almost everywhere. Hence, L_N/G_N can tend to a positive limit on a set of positive Lebesgue measure only if G_N is of the order of N. Without loss of generality we can therefore take $G_N = N$ (possibly turning to a subsequence of L_N and arguing for this subsequence only). Let us denote by $f(x)$ the limit of L_N/N as $N \to +\infty$, which is assumed to be positive on a set A. Assuming that $P(A) > 0$, we can introduce the measure $Q(B) = P(B|A)$. By Theorem 1,

$$\lim Q(L_N < \frac{Ny}{\log 2}) = \exp(-\frac{1}{y}), \quad (N \to +\infty).$$

The definition of $f(x)$ thus implies that $Q(f(x) < \frac{y}{\log 2}) = \exp(-\frac{1}{y})$. Put $D = \{f(x) < \frac{y}{\log 2}\}$. Applying Theorem 1 again with the measure $Q^*(B) = Q(B|D)$, we have, as $N \to +\infty$,

$$\text{(9)} \qquad \lim Q^*(L_N < \frac{Ny}{\log 2}) = \exp(-\frac{1}{y}) = Q(D).$$

The convergence of L_N/N to $f(x)$, however, yields that for $N \to +\infty$,

$$\text{(10)} \qquad \begin{aligned} \lim Q^*(L_N < \frac{Ny}{\log 2}) &= Q^*(f(x) < \frac{y}{\log 2}) \\ &= Q(D|D) = 1. \end{aligned}$$

Since (9) and (10) contradict each other, the assumption $P(A) > 0$ is wrong, i.e. $P(A) = 0$, completing the proof of Theorem 2.

3. The Remainders of the Oppenheim Expansion.

In this section I want only to make some comments on the order of magnitude of the remainders in the Oppenheim expansion of reals into infinite series of rationals. These expansions deserve greater attention since their speed of convergence is much faster than in the approximation by continued fractions, see the results in Galambos (1971). Let $a_n(j)$ and $b_n(j)$ be sequences of positive integer valued functions on the positive integers. The algorithm to obtain the infinite series representation

$$(11)\quad x = \frac{1}{d_1} + \frac{a_1}{b_1}\frac{1}{d_2} + \ldots + \frac{a_1 a_2 \ldots a_m}{b_1 b_2 \ldots b_m}\frac{1}{d_{m+1}} + \ldots$$

where $a_n = a_n(d_n)$ and $b_n = b_n(d_n)$ and $d_n = d_n(x)$, is as follows. Let $x = x_1$ and define the sequence $d_n(x)$ successively by

$$(12)\quad d_n = [1/x_n]+1\ , \quad x_n = 1/d_n+(a_n/b_n)x_{n+1}\ ,$$

where $[y]$ stands for the integer part of y. The infinite series representation (11) of x, together with the algorithm (12), is termed as the Oppenheim expansion of x, see Galambos (1970), Oppenheim (1972) and Schweiger (1972) for the related metric and number theoretic results. Put

$$h_n(j) = \{a_n(j)/b_n(j)\}j(j-1) .$$

It easily follows from (12), that

$$d_{n+1} > h_n(d_n) \quad \text{for all} \quad n ,$$

which cannot be sharpened to remain to hold for all x. Putting, however, R_N for the largest of the ratios $h_n(d_n)/d_{n+1}$, $1 \leq n \leq N$, it turns out that R_N tends to zero for almost all x. By a sieve argument, similar to the one employed in the previous section, the actual normalizing constants can be determined which are necessary for having a limit law. Note that the ratios above can be investigated through the sequence $y_n = h_n(d_n)x_{n+1}$, measuring the remainder in the expansion (11). Though the sequence y_n does not seem to be mixing, an almost equally powerful "almost independence" property is satisfied for these y's hence the major ideas to deal with this expansion remain the same as in the previous section, the actual details are, however, somewhat different, which will appear elsewhere. It is worth while pointing out that several classical expansions are special cases of the Oppenheim series. Specific choices of the functions $a_n(j)$ and $b_n(j)$ lead to the Engel, Sylvester and Lüroth series and to the Cantor products, as discussed in detail in Galambos (1970).

REFERENCES

[1] J. Galambos, The ergodic properties of the denominators in the Oppenheim expansion of real numbers into infinite series of rationals, Quart. J. Math. Oxford Ser. 21 (1970), 177-191.

[2] J. Galambos, On the speed of convergence of the Oppenheim series, Acta Arith. 19 (1971), 335-342.

[3] J. Galambos, The distribution of the largest coefficient in continued fraction expansions, Quart. J. Math. Oxford Ser. 23 (1972a), 147-151.

[4] J. Galambos, On the distribution of the maximum of random variables, Ann. Math. Statist., 43 (1972b), 516-521.

[5] A. Oppenheim, The representation of real numbers by infinite series of rationals, Acta Arith. 20 (1972), to appear.

[6] W. Philipp, Some metrical theorems in number theory II, Duke Math. J. 38 (1970), 447-458.

[7] A. Rényi, On stable sequences of events, Sankhya, Ser. A., 25 (1963), 293-302.

[8] F. Schweiger, Metrische Sätze über Oppenheimentwicklungen, J. Reine Angew. Math. 254(1972), 152-159.

Temple University, Philadelphia, Pa.

THE VALUES OF EXPONENTIAL POLYNOMIALS AT ALGEBRAIC POINTS (II)

Carlos J. Moreno

1. Introduction

In this paper we consider the values at algebraic points of the exponential polynomial

$$\phi(s) = \sum_{k=1}^{m} A_k \, e^{\alpha_k s},$$

where A_k, α_k, $(1 \leq k \leq m)$, are complex numbers which belong to an algebraic number field K of degree μ. The classical Theorem of Hermite-Lindemann states that if $\phi(s) \not\equiv 0$, then $\phi(\beta) \neq 0$ for any nonzero algebraic number β. Siegel [6] and Mahler [3] gave explicit lower bounds for $\phi(\beta)$ in terms of the size of the coefficients A_k. Using the ideas of Siegel, we obtained in [5] an explicit lower bound for $\phi(\beta)$ in terms of the height $H(\beta)$ of β. In this paper we investigate further the problem of obtaining a lower bound for $\phi(\beta)$ and obtain a significant improvement to the results in [5]. More precisely we obtain the following result.

THEOREM. Let $\phi(s)$ be the exponential polynomial

$$\phi(s) = \sum_{k=1}^{m} A_k e^{\alpha_k s},$$

where A_k, α_k, $(1 \leq k \leq m)$, are algebraic numbers. Let β be an algebraic number of height $H(\beta) = H$. Then

$$|\phi(\beta)| \geq \Lambda_1 \exp(-\Lambda_2 H^2) ,$$

where Λ_1 and Λ_2 are constants which depend only on the polynomial $\phi(s)$ and the degree of β.

As an immediate consequence of the above Theorem we obtain the following corollary.

COROLLARY. Let s_0 be a zero of the exponential polynomial $\phi(s)$. Let β be a nonzero algebraic number of height $H(\beta) = H$. Then

$$|s_0 - \beta| \geq \Lambda_1 \exp(-\Lambda_2 H^2) ,$$

where Λ_1 and Λ_2 are constants which depend only on the polynomial $\phi(s)$, the zero s_0 and the degree of β.

2. Notations.

Without loss of generality we may assume that the A_k, $(1 \leq k \leq m)$ are algebraic integers. We suppose further that $\omega_1, \ldots, \omega_\mu$ is a basis for the ring of integers of K and write the exponential polynomial $\phi(s)$ as

$$\phi(s) = \sum_{k_1=0}^{\sigma} \cdots \sum_{k_\mu=0}^{\sigma} A_{k_1 \ldots k_\mu} e^{(k_1\omega_1+\ldots+k_\mu\omega_\mu)s/d}$$

$$|A_{k_1 \ldots k_\mu}| \leqslant \max_{1 \leqslant k \leqslant m} |\overline{A_k}| = A,$$

where the coefficients $A_{k_1 \ldots k_\mu}$ are those of the original exponential polynomial suitably indexed, d is a rational integer and σ is a positive integer which depends only on the α_k, $(1 \leqslant k \leqslant m)$, and the choice of basis $\omega_1, \ldots, \omega_\mu$. The symbol $|\overline{\alpha}|$ means the largest of the absolute values of the conjugates of α. We may also assume that the α_k, $(1 \leqslant k \leqslant m)$, are algebraic integers in K, that is to say we may consider the polynomial $\phi^*(s) = \phi(ds)$. Then to obtain our result we consider the value of $\phi^*(s)$ at the algebraic point β/d.

3. Proof of Theorem.

We assume that no matter how large we take Λ, there always exist an algebraic number β of large height such that

$$\phi(\beta) << \exp(-\Lambda H^2) . \qquad (1)$$

In the spirit of Gelfond [1] Chap. II, §2 we shall subdivide the proof of the impossibility of (1) into four steps.

<u>Step One</u>. We enumerate the $q = (p+1)^{\mu}$ numbers

$$k_1\omega_1+\dots+k_\mu\omega_\mu, \qquad 0 \leqslant k_i \leqslant p_i, \qquad 1 \leqslant i \leqslant \mu,$$

where μ is a sufficiently large positive integer, in any order to form a sequence $\lambda_1,\dots,\lambda_q$ with the first $\xi = (\sigma+1)^{\mu}$ of them coinciding with the set of numbers $k_1\omega_1+\dots k_\mu\omega_\mu$, $0 \leqslant k_i \leqslant \sigma$, $i \leqslant i \leqslant \mu$. Thus we may write

$$\phi(\beta) = \sum_{k_1=0}^{\sigma} \dots \sum_{k_\mu=0}^{\sigma} A_{k_1\dots k_\mu} e^{(k_1\omega_1+\dots+k_\mu\omega_\mu)\beta}$$

as

$$\phi_1(\beta) = \sum_{k=1}^{q} A_{1,k} e^{\lambda_k \beta},$$

$$A_{1,k} = 0, k > \xi, \ |\overline{A_{1,k}}| \leqslant A .$$

If we multiply the inequality (1) by the numbers

$$e^{(k_1\omega_1+\dots+k_\mu\omega_\mu)\beta}, \qquad 0 \leqslant k_i \leqslant p-\sigma, \qquad 1 \leqslant i \leqslant \mu,$$

We shall obtain a set of inequalities of the type

$$(2) \qquad \phi_j(\beta) = \sum_{k=1}^{q} A_{j,k} e^{\lambda_k \beta} << \exp(-\Lambda H^2) ,$$

$$|\overline{A_{j,k}}| \leqslant A, \quad 1 \leqslant j \leqslant (p+1-\sigma)^{\mu} = \tau,$$

where the constant Λ now depends on p. In fact p will be a parameter which will be chosen later to be not larger than some power of log H.

The linear independence of the ω_i, $(1 \leq j \leq \mu)$, implies that the q by τ matrix $(A_{j,k})$, $(1 \leq j \leq \tau,\ 1 \leq k \leq q)$, formed with the coefficients of the linear forms $\phi_j(\beta)$ has rank τ. The next step consists in producing a new set of q independent linear forms in the exponentials $e^{\lambda_k \beta}$ from which a set of $\eta = q-\tau$ forms can be selected so as to complete the system in (2).

<u>Step Two</u>. The construction of q independent linear forms in the exponentials $e^{\lambda_k \beta}$, $(1 \leq k \leq q)$ was first done by Siegel in [6]. Here we follow Mahler [3] who gave a simplified version of the method. The presentation is that of Leveque [2], where more details can be found.

Let $r_1,\ldots,r_q$ be q positive integers and $\lambda_1,\ldots,\lambda_q$ the complex numbers considered above. Then one can construct a unique set of polynomials

$$A_k(s) = A_k(s;\ r_1,\ldots,r_q;\ \lambda_1,\ldots,\lambda_q),\ 1 \leq k \leq q\ ,$$

where the k-th polynomial has degree r_k, such that the function

$$\begin{aligned} R(s) &= R(s;\ r_1,\ldots,r_q;\ \lambda_1,\ldots,\lambda_q)\ , \\ &= A_1(s)e^{\lambda_1 s}+\ldots+A_q(s)e^{\lambda_q s} \end{aligned}$$

has a MacLaurin expansion about $s = 0$

$$R(s) = \frac{s^{r-1}}{(r-1)!} + b_r s^r + \ldots\ ,$$

where $r = r_1+\dots+r_q$. In fact it can be shown that $R(s)$ has an integral representation of the form

(3) $$R(s) = \int_0^s dt_{q-1} \int_0^{t_{q-1}} dt_{q-2} \cdots$$
$$\times \int_0^{t_2} \frac{t_1^{r_1-1}(t_2-t_1)^{r_2-1}\cdots(t_{q-1}-t_{q-2})^{r_{q-1}-1}(s-t_{q-1})^{r_q-1}}{(r_1-1)!\cdots(r_q-1)!}$$
$$\times \exp\{\lambda_1 t_1+\lambda_2(t_2-t_1)+\dots+\lambda_q(s-t_{q-1})\}dt_1 .$$

For h and k in the sequence $1,2,\dots,q$, let

$$\delta_{hk} = \begin{cases} 1 & \text{if } h=k \\ 0 & \text{if } h\neq k \end{cases}$$

For a fixed but arbitrary positive integer ρ, we put

$$R_j(s) = R_j(s;\rho;\lambda_1,\dots,\lambda_q)$$
$$= R(s;\rho+\delta_{1j},\dots,\rho+\delta_{qj};\lambda_1,\dots,\lambda_q).$$

and

$$A_{jk}(s) = A_{jk}(s;\rho;\lambda_1,\dots,\lambda_q)$$
$$= A_k(s;\rho+\delta_{1j},\dots,\rho+\delta_{qj};$$
$$\lambda_1,\dots,\lambda_q) .$$

The system of polynomials $A_{hk}(s)$ possesses the fundamental property that the square matrix

$$A(s) = (A_{jk}(s)), \quad (1 \leqslant j, \quad k \leqslant q)$$

has determinant

$$D(S) = s^{q\rho}(\rho!)^{-q} \prod_{k=1}^{q} \prod_{\substack{\ell=1 \\ \ell\neq 1}}^{q} (\lambda_k-\lambda_\ell)^{-\rho} .$$

Clearly $D(\beta) \neq 0$ if $\beta \neq 0$. Hence the system of q linear forms

$$R_j(\beta) = A_{j1}(\beta)e^{\lambda_1\beta}+\ldots+A_{jq}(\beta)e^{\lambda_q\beta} , \quad (1 \leqslant j \leqslant q)$$

is complete.

Now we would like to find a common denominator for the coefficients of the polynomials $A_{jk}(s)$. To accomplish this we recall the explicit formula given in [2] pg. 178 for the polynomials $A_{jk}(s)$ and which are obtained by formally expanding the operator identity

$$A_{hk}(s) = \left\{ \prod_{\substack{j=1 \\ j\neq k}}^{q} (D+\omega_k-\omega_j)^{-\rho^*+1} \right\} s^{\rho^*}/\rho^*! ,$$

where $D = d/ds$ and $\rho^* = \rho+\delta_{hk}-1$. Furthermore from the operator identity

$$(\rho-1)!(1-zD)^{-\rho} = \sum_{v=0}^{\infty} (v+\rho-1)\ldots(v+1)z^v D^v$$

we obtain

$$A_{hk}(s) = (1/\rho^*!) \prod_{\substack{j=1 \\ j\neq k}}^{q} (\lambda_k-\lambda_j)^{-2\rho-\delta_{jh}}$$

(4)

$$\times \sum_{w=0}^{\rho^*} \left\{ \sum_{\substack{v_1=0 \\ v_1+\dots+\hat{v}_k+\dots+v_q=w}}^{\rho^*} \cdots \sum_{v_q=0}^{\rho^*} \prod_{\substack{j=1 \\ j\neq k}}^{q} \binom{v_j+\rho^*}{\rho^*} (\lambda_k-\lambda_h)^{-\rho-v_j} \right\}$$

$$(-1)^w D^w s^{\rho^*} ,$$

where $\binom{m}{n}$ is the n-th coefficient in the expansion $(1+x)^m$ and the symbol $\hat{v}_k$ is written to mean that the summation corresponding to v_k is omitted.

From (4) it follows that a common denominator for all the polynomials $A_{kh}(s)$ is given by

$$\rho!\Omega^{2\rho+1},$$

where

$$\Omega = \prod_{1\leq h<k\leq q} (\lambda_k-\lambda_h) .$$

Let us put

$$(5) \quad \rho!\Omega^{2\rho+1} R_h(\beta) = a_{h1}(\beta)e^{\lambda_1\beta}+\dots+a_{nq}(\beta)e^{\lambda_q\beta} ,$$

where the $a_{hk}(\beta)$ are polynomials in β with coefficients which are algebraic integers. Now consider the set of $\tau = (p+1-\sigma)^\mu$ forms in (2), all of which are linearly independent, and complete the system by selecting $\eta = q-\tau$ forms from (5), say

$$\psi_h(\beta) = a_{h1}(\beta)e^{\lambda_1\beta}+\dots+a_{hq}(\beta)e^{\lambda_2\beta} , \ 1 \leq h \leq \eta ,$$

so that the resulting matrix of coefficients

$$M = \begin{pmatrix} A_{11} & \cdots & A_{1q} \\ \vdots & & \vdots \\ A_{\tau 1} & \cdots & A_{\tau q} \\ & & \\ a_{11}(\beta) & \cdots & a_{1q}(\beta) \\ \vdots & & \vdots \\ a_{\eta 1}(\beta) & \cdots & a_{\eta q}(\beta) \end{pmatrix}$$

has rank $q = \tau+\eta$, i.e., $\det(M)$ is a nonzero algebraic number with a denominator which depends only on β and ρ.

Step Three. The essential idea of the proof consists in observing that on the one hand the determinant of the matrix M can be made small by suitable choices of the parameters p and ρ and using the inequalities (2). On the other hand, $\det(M)$ is a nonzero algebraic number which after clearing denominators has a norm which is in absolute value larger than 1. This then leads to a lower bound for $\det(M)$ which cannot be too small. To finish the proof, one observes that a balanced choice of the parameters p and ρ as functions of $H = H(\beta)$ leads to a contradiction.

The determinant of the matrix M is a polynomial in β of degree $\eta\rho$ with coefficients which are algebraic integers in the field K of degree μ. Therefore the number

$$D = H^{\eta\rho} \det(M)$$

is an algebraic integer in the field $K(\beta)$ of degree μd over the rationals and hence it has a norm greater than 1, i.e.

$$|\text{Norm } (D)| \geq 1 .$$

Hence to get a lower bound for D it is enough to get upper bounds for the conjugates of D. For this computation we need estimates for the algebraic size of the coefficients of the polynomials $a_{hk}(\beta)$.

We observe first that

$$a_{hk}(s) = \rho!\Omega^{2\rho+1} A_{hk}(s) .$$

Hence (4), (5) and the following Lemmas give the desired estimates for the conjugates of D.

LEMMA 1. We have for $k \leq q$

$$(6) \qquad \rho!\Omega^{2\rho+1}(\rho^*!)^{-1} \prod_{\substack{j=1 \\ j\neq k}}^{q} (\lambda_k - \lambda_j)^{-2\rho-\delta_{jh}}$$

$$<< e^{4\rho q^2 \log p} ,$$

with ρ larger than some constant c which depends only on $\omega_1, \ldots, \omega_\mu$.

Proof. The estimate (6) is essentially a trivial one and follows from

$$|\lambda_k-\lambda_j| \leqslant \mu p \operatorname{Max}(|\omega_j|) .$$

LEMMA 2. We have for $w \leqslant \rho$

$$(7) \qquad \sum_{\substack{v_1=0 \\ v_1+\dots+\hat{v}_k+\dots+v_q=w}}^{\rho^*} \dots \sum_{v_q=0}^{\rho^*} \prod_{\substack{j=1 \\ j\neq k}}^{q} \binom{\rho^*+v_j}{\rho^*}(\lambda_k-\lambda_h)^{\rho-v_j}$$

$$<< e^{3q\rho \log p} ,$$

where $\rho^* = \rho+\delta_{hk}-1$, and ρ is larger than some constant which depends only on $\omega_1,\dots,\omega_\mu$.

Proof. The number of terms in the sum (7)

$$\leqslant \rho^q \leqslant e^{q \log \rho} .$$

the size of each term is bounded by

$$\prod_{\substack{j=1 \\ j\neq k}}^{q} 2^{v_j+\rho}(p\mu \operatorname{Max}(|\omega_j|))^{\rho-v_j} << e^{2\rho q \log p} ,$$

for ρ larger than some constant c which depends only on $\omega_1,\dots,\omega_\mu$. Hence the Lemma

LEMMA 3. We have for $w \leqslant \rho$

$$(8) \qquad |D^w s^{\rho^*}| << e^{\rho \log \rho}|s|^{\rho^*-w}$$

Proof. The above Lemma follows easily from the identity

$$D^w s^{\rho^*} = (\rho^*)\dots(\rho^* + 1-w)s^{\rho^*-w} .$$

From (4)-(8) it is clear that the coefficients of the polynomials $a_{kh}(s)$ have all their conjugates bounded by

$$(9) \qquad \ll e^{5\rho q^2 \log p + \rho \log \rho}$$

Finally to find an upper bound for any of the conjugates of the algebraic numbers $a_{hk}(\beta)$ we simply observe that the conjugates of β are all bounded by

$$dH(\beta) ,$$

where d and $H = H(\beta)$ are the degree and height of β respectively. This remark leads to the following Lemma.

LEMMA 4. Any conjugate in $K(\beta)$ of the algebraic number $a_{hk}(\beta)$ is bounded by

$$(10) \qquad \ll e^{5\rho q^2 \log p}(dH)^{\rho} .$$

Proof. (10) is a direct consequence of (9) and the fact that any embedding of the field $K(\beta)$, where the numbers $a_{hk}(\beta)$ lie, is induced by extending the embeddings of K and $Q(\beta)$.

Upper bounds for any of the conjugates of the determinant of the matrix M are obtained by using the expansion

$$\det(m) = \sum_{\pi} (\pm 1) A_{1\pi(1)} \cdots A_{\tau\pi(\tau)} a_{1\pi(\tau+1)} \cdots a_{\eta\pi(\tau+\eta)} ,$$

where the sum runs over all permutations π of the set $(1,2,\ldots,q)$, and observing that

$$\det(M) \leq q!(\mathrm{Max}|A_{jk}|)^{\tau}(\mathrm{Max}|a_{hk}(\beta)|)^{\eta} .$$

Hence we have

$$i(D) << A^{\tau} H^{2\rho\eta} e^{6\eta\rho q^2 \log p + \eta\rho \log \rho} ,$$

where the symbol $i(D)$ means conjugation. We have also observed that the norm of D is an integer having absolute value greater than 1. Hence

$$1 \leq |\prod_i i(D)| << |D| A^{(d\mu-1)} H^{2(d\mu-1)\eta\rho} e^{(d\mu-1)(6\eta\rho q^2 \log p + \eta\rho \log \rho)} .$$

Thus we have the following basic lower bound

$$(11) \qquad |D| >> A^{-c_1\tau} H^{-c_2\eta\rho} e^{-c_3(6\eta\rho q^2 \log p + \eta\rho \log \rho)} ,$$

where the constants c_1, c_2, and c_3 depend only on $\omega_1, \ldots, \omega_\mu$ and d.

To obtain upper bounds for $\det(D)$ we use on the first τ rows the inequalities (2) and on the last η rows we use upper bounds for the numbers

$$\psi_h(\beta) = \rho!\Omega^{2\rho+1} R_h(\beta) .$$

To estimate the numbers $R_h(\beta)$ we use the integral representation given by (3) to obtain

$$|R_h(\beta)| \ll |\beta|^{q\rho}((\rho-1)!)^{-q} e^{c_4 q p|\beta|} ,$$

and recall that

$$\rho!\Omega^{2\rho+1} \ll e^{4\rho q^2 \log p + \rho \log \rho} .$$

Therefore we have

$$(12) \qquad \psi_h(\beta) \ll \exp\{5\rho q^2 \log p + c_4 q p|\beta| + q\rho \log H - (q-1)\rho \log \rho\} .$$

Let us now multiply all the elements in the first column of the matrix M by the number $e^{\lambda_1 \beta}$. Add to the resulting column the j-th column multiplied by the number $e^{\lambda_j \beta}$ for all $j = 2,\dots,q$. The new matrix still has determinant identical to $\det(M)$ but now looks like

$$\begin{pmatrix} \phi_1(\beta) & A_{12} & \cdots & A_{1q} \\ \vdots & & & \\ \phi_\tau(\beta) & A_{\tau 2} & \cdots & A_{\tau q} \\ \psi_1(\beta) & a_{12}(\beta) & \cdots & a_{1q}(\beta) \\ \vdots & & & \\ \psi_\eta(\beta) & a_{\eta 2}(\beta) & \cdots & a_{\eta q}(\beta) \end{pmatrix}$$

The determinant of this matrix can be written as

$$\det(M) = \sum_{\pi} (\pm 1) A_{1\pi(1)} \cdots \hat{A}_{j\pi(j)} \cdots A_{\tau\pi(\tau)} a_{1\pi(\tau+1)}(\beta) \cdots a_{\eta\pi(q)}(\beta)$$

$$+ \sum_{\pi}{}'' (\pm 1) A_{1\pi(1)} \cdots A_{\tau\pi(\tau)} a_{1\pi(\tau+1)}(\beta) \cdots \hat{a}_{j\pi(\tau+j)}(\beta) \cdots a_{\eta\pi(q)}(\beta) \ ,$$

where in the first sum $\sum$ the symbol $\hat{A}_{j\pi(j)}$ means that some $\phi_j(\beta)$ occupies that place; similarly in the second sum $\sum''$ the symbol $\hat{a}_{j\pi(\tau+j)}(\beta)$ means that some $\psi_h(\beta)$ occupies that place. Taking absolute values we get

$$|\det(M)| \leq q!\{A^{\tau-1}(\mathrm{Max}|\phi_j(\beta)|)(\mathrm{Max}|a_{hk}(\beta)|)^{\eta} + A^{\tau}(\mathrm{Max}|\psi_j(\beta)|)(\mathrm{Max}|a_{hk}(\beta)|)^{\eta-1}\}$$

Therefore

$$|D| \leq q! H^{\eta\rho} A^{\tau-1}(\mathrm{Max}|a_{hk}(\beta)|)^{\eta-1} \ .$$
$$\{(\mathrm{Max}|\phi_j(\beta)|)(\mathrm{Max}|a_{kh}(\beta)|) + A(\mathrm{Max}|\psi_j(\beta)|)\} \ .$$

From the inequalities (2), (10) and (12) we have

(13) $$|D| << H^{c_5\eta\rho} A^{\tau} e^{c_6\eta\rho q^2 \log p + c_7\eta\rho\log\rho}$$

$$\{\exp(-\Lambda H^2) + \exp(c_4 qp(\beta) + q\rho\log H - q\rho\log\rho)\} \ .$$

From (11) and (13) we have

(14)

$$1 << A^{c_9\tau}(H^{c_8}\exp(c_6q^2\log p + c_7 \log \rho))^{\eta\rho}$$

$$\{\exp(-\Lambda H^2) + \exp(c_{11}pqH + q\rho\log H - q\rho\log \rho)\} .$$

Step Four. The final step consists in balancing our selection of ρ and q as functions of H so that the inequality (14) is impossible. We begin by putting $\rho = H^{3/2}$ so that

$$c_{11}pq\ H + q\rho\log H - q\rho \log \rho$$

$$= c_{11}q\nu\rho^{2/3} - (q\rho \log\rho)/3$$

We now want to select q as a function of ρ so that

$$c_6\eta\rho q^2\log p = (q\rho \log \rho)/8 .$$

This can clearly be done by putting

$$q = c_{14}((\log \rho)/\log \log p)^{\mu/(2\mu-1)}$$

with a suitable constant c_{14}. Simply observe that

$$\eta = q-(p+1-\sigma)^\mu = (p+1)^\mu-(p+1-\sigma)^\mu \leqslant \mu\sigma q^{(\mu-1)/\mu}$$

and

$$\log p = c_{15}\log q + O(1)$$

$$= c_{16}\log \log \rho + O(\log \log \log \rho) .$$

Furthermore since $p = q^{1/\mu}-1$, we have

$$c_{11}pg\ H \leq (q\rho \log \rho)/8$$

$$c_7\eta\rho \log \rho \leq (q\rho \log \rho)/8$$

$$c_8\eta\rho \log H \leq (q\rho \log \rho)/8$$

$$c_9\tau \log A \leq (q\rho \log \rho)/8$$

Hence from the inequality (14) we obtain

$$1 << (\exp\{-\Lambda H^2 + c_{15}H^{5/3}\}$$

$$+ \exp\{-(q\rho \log \rho)/10\}) ,$$

and clearly this is a contradiction if H is large. Therefore the assumption that the inequalit (1) holds is wrong and hence

$$\phi(\beta) >> \exp(-\Lambda H^2)$$

which is what we wanted to prove. This completes the proof of the Theorem.

4. Remarks.

It is clear from the above proof that the exponent in H^2 can be slightly improved. What is really desirable is to be able to replace the lower bound $\exp(-\Lambda H^2)$ by $\exp(-\Lambda H^\gamma)$ with $\gamma < 1$ and preferable by $\exp(-\Lambda \log H)$.

Results of Mahler concerning the rational approximations of π show that indeed a polynomial bound is possible for the value of certain exponential polynomials at an algebraic point. In Mahler's case one deals with the simple polynomial

$$\phi(s) = e^{is} + 1 .$$

A close analysis of the above proof and the results of Mahler [4] suggest that a significant improvement can be obtained if one can estimate the expression

$$\Omega = \prod_{1 \leq h < k \leq q} (\lambda_k - \lambda_h)$$

in a nontrivial manner.

REFERENCES

[1] A.O. Gelfond, Transcendental and Algebraic Numbers, Dover, New York, 1960.
[2] W. Leveque, Topics in Number Theory, Vol. 2, Addison-Wesley, Reading, 1956.
[3] K. Mahler, Zur approximation der exponential funktion und des logarithmus, Journal f.d. reine u. angew. Math., 166 (1932).
[4] K. Mahler, On the approximation of π, Proceedings, Koninklijke Nederl. Akad. Wetensch., Amsterdam, Ser. A56 (1953).
[5] C. Moreno, The value of exponential polynomials at algebraic points (I), to appear.
[6] C.L. Siegel, Uber einige anwendunger diophantischer approximationen, Abh. Preuss. Akad. Wiss., 1929-1930, No. 1.

The Center for Advanced Study, University of Illinois, Urbana, Illinois

APPLICATION OF DIOPHANTINE APPROXIMATIONS TO NUMERICAL INTEGRATION

H. Niederreiter

0. Introduction

The last two decades, which have seen the development of the Monte Carlo method into an effective and widely accepted tool for numerical integration, have also brought us, in a somewhat parallel evolution, the emergence of certain deterministic integration methods which could be collectively called number-theoretic integration methods. In this paper, I shall be concerned with two such methods, which I label the method of low discrepancy sequences and the Fourier series method respectively. Both will depend heavily on techniques and results from the theory of diophantine approximations. A third method, the so-called method of good lattice points (or optimal coefficients), will be discussed in detail at this conference by S.K. Zaremba.

The basic problem of numerical integration is of course the approximate computation of definite integrals. In the one-dimensional case, we can always assume that the given integral $\int_a^b g(u)du$ is transformed, by means of a suitable change of variable, into an integral of the form $\int_0^1 f(t)dt$. In the multidimensional case, we can assume that, again after a suitable change of variables, the integration domain is a subset of a multidimensional unit cube. The emphasis will be on the case where this unit cube itself is used as the integration domain. However, in section 7 we shall also study a very general class of integration domains. Almost all our results will deal with equal-weight formulas, with an important deviation occurring in section 9.

We shall establish a good number of new results pertaining to both integration methods to be described here. Moreover, we will include the proofs of some known results for which a short proof can be given, or where we felt that our proof is much simpler than the ones available in the literature. The paper is organized as follows:

Section 1 sets the stage for the method of low discrepancy sequences in the one-dimensional case. We introduce the two most common notions of discrepancy, and review briefly some of their important properties.

The central theme in section 2 is the estimation of the integration error for equal-weight formulas in terms of discrepancy in the one-dimensional case. All estimates require only a low degree of regularity of the integrand, and most of them are easy to prove.

The success of the method of low discrepancy sequences depends on our ability to exhibit infinite sequences with uniformly small discrepancy. In section 3 we show that the theory of diophantine approximations can be most instrumental in providing such sequences. We present discrepancy estimates for the sequences $(\{n\alpha\})$, $n = 1,2,\ldots,$ α irrational, based either on the Erdős-Turán inequality or on continued fractions.

In sections 4, 5 and 6, the results of the previous sections are extended, as far as possible, to the multidimensional case. There are still many open problems in this area.

In section 7, we show how the method of low discrepancy sequences can also be used effectively for any integration domain which is a continuity set with respect to Lebesgue measure. A whole new family of discrepancies is introduced in this connection.

The second part of the paper is devoted to the Fourier series method. In section 8, we describe this method, and we also compare the efficiency of the two methods with respect to various integration problems.

An important refinement of the Fourier series method is presented in section 9. We show that, by using suitable weight distributions, we can arrive at methods which are rapidly convergent for integrands with a high degree of regularity.

Finally, in section 10, we exhibit the intimate connection between the problem of explicit error bounds in these number-theoretic integration methods and the so-called effective methods in diophantine approximations. There is an abundance of open problems here, most of which seem to be extremely difficult.

For surveys of number-theoretic integration methods, we refer to the following books and articles: Davis and Rabinowitz [11,Ch.5], Haber [15], Halton [17], [18], Hammersley and Handscomb [20, Ch. 3], Hlawka [29], Korobov [35], Kuipers and Niederreiter [37, section 2.5], and Zaremba [69].

Part A. The Method of Low Discrepancy Sequences.

1. Discrepancy

Let $E = [0,1]$ be the closed unit interval. Given a finite sequence $x_1,\ldots,x_N$ of points in E, we define the remainder function

$$R_N(t) = \frac{A(t;N)}{N} - t \quad \text{for} \quad 0 \leqslant t \leqslant 1 ,$$

where $A(t;N)$ denotes the number of elements x_n, $1 \leqslant n \leqslant N$, which are less than t. We shall be concerned with the following two notions of discrepancy (for other notions, see [45]).

Definition 1.1. The (extreme) discrepancy D_N of the sequence $x_1,\ldots,x_N$ in E is defined to be

$$D_N = \sup_{0 \leq t \leq 1} |R_N(t)| .$$

Definition 1.2. The L^2 discrepancy $D_N^{(2)}$ of the sequence $x_1, \ldots, x_N$ in E is defined to be

$$D_N^{(2)} = \left(\int_0^1 R_N{}^2(t)dt\right)^{1/2} .$$

Definition 1.3. Let $\omega = (x_n)$, $n = 1,2,\ldots,$ be an infinite sequence in E, and let N be a positive integer. Then $D_N(\omega)$, resp. $D_N^{(2)}(\omega)$, stands for the extreme, resp. L^2, discrepancy of the first N terms of ω.

There are easy methods available for computing these discrepancies. For a more systematic pursuit of this aspect, we refer to Niederreiter [43]. If we arrange the elements of a finite sequence according to their magnitude, then the extreme discrepancy is obtained readily from the following formula.

THEOREM 1.1 (Niederreiter [42]). If $0 \leq x_1 \leq x_2 \leq \ldots \leq x_N \leq 1$, then the discrepancy D_N of $x_1, \ldots, x_N$ is given by

$$D_N = \frac{1}{2N} + \max_{1 \leq n \leq N} \left|x_n - \frac{2n-1}{2N}\right| . \tag{1.1}$$

For a generalization of this formula, see Niederreiter [45]. We may deduce from (1.1) that $D_N \geq \frac{1}{2N}$ for any sequence of N elements in E (a simple fact already known to van der Corput and Pisot [60] and that $D_N = \frac{1}{2N}$ exactly for the sequence $\frac{1}{2N}, \frac{3}{2N}, \ldots, \frac{2N-1}{2N}$ or one of its rearrangements. If we consider infinite sequences, then deeper results concerning lower bounds for D_N are known. Schmidt [55] has shown that for any infinite sequence ω in E there exist infinitely many N such that $D_N(\omega) \geqslant (100\ N)^{-1}$ $\log N$. On the other hand, we will see in section 3 that there are sequences ω with $ND_N(\omega) = O(\log N)$.

For $D_N^{(2)}$ two types of formulas can be shown. One is fairly explicit, and the other one is an identity useful for the estimation of $D_N^{(2)}$.

<u>THEOREM 1.2.</u> (Koksma [33]) For any $x_1, \ldots, x_N$ in E we have

$$(D_N^{(2)})^2 = \frac{1}{3} + \frac{1}{N^2} \sum_{n=1}^{N} (Nx_n{}^2 + x_n)$$
$$- \frac{2}{N^2} \sum_{n=1}^{N} \sum_{m=1}^{n} \max(x_n, x_m) .$$

<u>Proof</u>. Define $c(t,x) = 1$ if $x < t$, and $c(t,x) = 0$ otherwise. Then

$$(ND_N^{(2)})^2 = \int_0^1 \Big(\sum_{n=1}^{N} c(t,x_n) - Nt \Big)^2 dt$$
$$= N^2 \int_0^1 t^2 dt - 2N \sum_{n=1}^{N} \int_0^1 tc(t,x_n) dt$$

$$+ \sum_{n=1}^{N} \sum_{m=1}^{N} \int_0^1 c(t,x_n)c(t,x_m)dt$$

$$= \frac{1}{3}N^2 - N\sum_{n=1}^{N}(1-x_n^2) + \sum_{n=1}^{N}\sum_{m=1}^{N}(1-\max(x_n,x_m))$$

$$= \frac{1}{3}N^2 + \sum_{n=1}^{N}(Nx_n^2+x_n) - 2\sum_{n=1}^{N}\sum_{m=1}^{n}\max(x_n,x_m) .$$

COROLLARY 1.1. If $0 \leqslant x_1 \leqslant x_2 \leqslant \ldots \leqslant x_N \leqslant 1$, then the L^2 discrepancy $D_N^{(2)^2}$ of $x_1,\ldots,x_N$ is given by

$$(1.2) \quad (D_N^{(2)})^2 = \frac{1}{N}\sum_{n=1}^{N}\left(x_n - \frac{2n-1}{2N}\right)^2 + \frac{1}{12N^2} .$$

Proof. This follows from Theorem 1.2 by a straightforward computation.

The identity (1.2) implies that $D_N^{(2)} \geqslant \frac{1}{\sqrt{12}N}$ for any N elements in E, and that $D_N^{(2)} = \frac{1}{\sqrt{12}N}$ exactly for the sequence $\frac{1}{2N}, \frac{3}{2N}, \ldots, \frac{2N-1}{2N}$ or one of its rearrangements. The following is an unpublished result of Koksma (see also Kuipers [36]).

<u>THEOREM 1.3.</u> For any $x_1,\ldots,x_N$ in E we have

$$(1.3)\qquad (D_N^{(2)})^2 = \Big(\frac{1}{N}\sum_{n=1}^{N}(x_n-\tfrac{1}{2})\Big)^2 + \frac{1}{2\pi^2}\sum_{h=1}^{\infty}\frac{1}{h^2}\Big|\frac{1}{N}\sum_{n=1}^{N}e^{2\pi i h x_n}\Big|^2 .$$

<u>Proof.</u> The function $R_N(t)$, being piecewise linear in E, can be expanded into a Fourier series $\sum_{h=-\infty}^{\infty} a_h e^{2\pi i h x}$ which will represent $R_N(t)$ apart from finitely many points. Using the function $c(t,x)$ from the proof of Theorem 1.2, we obtain

$$a_0 = \int_0^1 R_N(t)\,dt = \frac{1}{N}\sum_{n=1}^{N}\int_0^1 c(t,x_n)\,dt - \int_0^1 t\,dt$$

$$= \frac{1}{N}\sum_{n=1}^{N}(1-x_n) - \frac{1}{2} = -\frac{1}{N}\sum_{n=1}^{N}(x_n-\tfrac{1}{2}) .$$

For $h \neq 0$, we have

$$a_h = \int_0^1 R_N(t)e^{-2\pi i h t}\,dt$$

$$= \frac{1}{N}\sum_{n=1}^{N}\int_0^1 c(t,x_n)e^{-2\pi i h t}\,dt - \int_0^1 t e^{-2\pi i h t}\,dt$$

$$= \frac{1}{2\pi i h N}\sum_{n=1}^{N}(e^{-2\pi i h x_n}-1) + \frac{1}{2\pi i h}$$

$$= \frac{1}{2\pi i h N}\sum_{n=1}^{N}e^{-2\pi i h x_n} .$$

By Parseval's identity, we have

$$\int_0^1 R_N^2(t)dt = a_0^2 + 2\sum_{h=1}^{\infty} |a_h|^2 ,$$

and the desired result follows.

COROLLARY 1.2. For any N points in E we have

$$\frac{1}{\sqrt{12}} D_N^{3/2} \leqslant D_N^{(2)} \leqslant D_N .$$

Proof. The second inequality is obvious from the definitions. The first inequality follows from Theorem 1.3 and the well-known inequality of LeVeque [38] to the effect that

$$D_N \leqslant \left(\frac{6}{\pi^2}\sum_{h=1}^{\infty}\frac{1}{h^2}\left|\frac{1}{N}\sum_{n=1}^{N} e^{2\pi i h x_n}\right|^2\right)^{1/3} .$$

There is an important estimate for D_N in terms of the exponential sums occurring in (1.3). We shall need this inequality in section 3.

THEOREM 1.4. (Erdős and Turán [12]). For any $x_1,\ldots,x_N$ in E and for any positive integer m, the discrepancy D_N of $x_1,\ldots,x_N$ satisfies

$$D_N \leqslant \frac{6}{m} + \left(\frac{4}{\pi}\right)\sum_{h=1}^{m}\frac{1}{h}\left|\frac{1}{N}\sum_{n=1}^{N} e^{2\pi i h x_n}\right| . \tag{1.4}$$

Erdős and Turán proved (1.4) with unspecified constants. The validity of (1.4) with the specific constants is established in a forthcoming paper of Niederreiter and

Philipp (see also [37, ch. 2]). Another proof of (1.4) but with much larger constants was provided by Yudin [63].

2. Error Estimates

It is well known that an infinite sequence $\omega = (x_n)$ in E is uniformly distributed in E (in the sense of Weyl [62]) if and only if $\lim_{N\to\infty} D_N(\omega) = 0$. This is in fact just a special case of the Pólya-Cantelli Theorem on pointwise convergence of monotone functions ([13, pp. 319-321]). It follows then from Corollary 1.2 that ω is uniformly distributed in E if and only if $\lim_{N\to\infty} D_N^{(2)}(\omega) = 0$ (see also Niederreiter [45] for a direct proof). If ω is uniformly distributed in E, then by Weyl's criterion we must have

$$(2.1) \qquad \lim_{N\to\infty} \frac{1}{N} \sum_{n=1}^{N} f(x_n) = \int_0^1 f(t)\,dt$$

for every Riemann integrable function f on E. This observation is the basis of our integration method. One asset of this method is that we can establish satisfactory error estimates even for integrands which are not very smooth. It is a common feature of these error estimates that the speed of convergence in (2.1) is set in a direct relation with the order of magnitude of $D_N(\omega)$ or $D_N^{(2)}(\omega)$.

LEMMA 2.1. Let f be a continuous function on E of bounded variation. Then the identity

$$\int_0^1 f(t)dt - \frac{1}{N}\sum_{n=1}^{N} f(x_n) = \int_0^1 R_N(t)df(t)$$

holds for any $x_1,\ldots,x_N$ in E.

Proof. Let $c(t,x)$ be the function introduced in the proof of Theorem 1.2. Then

$$\int_0^1 R_N(t)df(t) = \frac{1}{N}\sum_{n=1}^{N}\int_0^1 c(t,x_n)df(t) - \int_0^1 t\ df(t) .$$

Using integration by parts for the last integral, we infer that

$$\begin{aligned}\int_0^1 R_N(t)df(t) &= \frac{1}{N}\sum_{n=1}^{N}(f(1)-f(x_n))-f(1) + \int_0^1 f(t)dt\\ &= -\frac{1}{N}\sum_{n=1}^{N} f(x_n) + \int_0^1 f(t)dt.\end{aligned}$$

THEOREM 2.1. (Koksma [32]). Let f be a continuous function on E of bounded variation $V(f)$, and let $x_1,\ldots,x_N$ be points in E with discrepancy D_N. Then

$$\left|\frac{1}{N}\sum_{n=1}^{N} f(x_n) - \int_0^1 f(t)dt\right| \leq V(f)D_N .$$

Proof. This follows immediately from Definition 1.1 and Lemma 2.1.

Koksma's inequality holds in fact for all functions f on E of bounded variation, but the proof of the general case is more elaborate.

Various extensions of Koksma's inequality pertinent to other numerical integration problems are known (Helmberg [25], Hlawka and Mück [31]).

THEOREM 2.2. (Sobol' [56], Zaremba [66]). Suppose f has a continuous derivative on E, and let $x_1,\ldots,x_N$ be points in E with L^2 discrepancy $D_N^{(2)}$. Then

$$\left|\frac{1}{N}\sum_{n=1}^{N} f(x_n) - \int_0^1 f(t)dt\right| \leq D_N^{(2)} \left(\int_0^1 (f'(t))^2 dt\right)^{1/2}.$$

Proof. From Lemma 2.1, we deduce the identity

$$\left|\frac{1}{N}\sum_{n=1}^{N} f(x_n) - \int_0^1 f(t)dt\right| = \left|\int_0^1 R_N(t)f'(t)dt\right|.$$

An application of the Cauchy-Schwarz inequality to the right hand side completes the proof.

THEOREM 2.3. (Niederreiter [43]). Let f be continuous on E with modulus of continuity M, and let $x_1,\ldots,x_N$ be points in E with discrepancy D_N. Then

$$\left|\frac{1}{N}\sum_{n=1}^{N} f(x_n) - \int_0^1 f(t)\,dt\right| \leqslant M(D_N) .$$

Proof. Without loss of generality, we may assume $x_1 \leqslant x_2 \leqslant \ldots \leqslant x_N$. We write

$$\int_0^1 f(t)\,dt = \sum_{n=1}^{N} \int_{\frac{n-1}{N}}^{\frac{n}{N}} f(t)\,dt = \frac{1}{N}\sum_{n=1}^{N} f(\xi_n)$$

with $\frac{n-1}{N} < \xi_n < \frac{n}{N}$ for $1 \leqslant n \leqslant N$. Therefore

$$\frac{1}{N}\sum_{n=1}^{N} f(x_n) - \int_0^1 f(t)\,dt = \frac{1}{N}\sum_{n=1}^{N} (f(x_n)-f(\xi_n)).$$

From Theorem 1.1 and the condition on the ξ_n it follows that $|x_n-\xi_n| \leqslant D_N$ for $1 \leqslant n \leqslant N$. But then $|f(x_n) - f(\xi_n)| \leqslant M(D_N)$ for $1 \leqslant n \leqslant N$, and we are done.

For a partition $\mathcal{P}$ of $[0,1)$ into half-open intervals, let $n(\mathcal{P})$ be the maximal length of the intervals in $\mathcal{P}$. If f is a bounded function on E, then $s(f;\mathcal{P})$ shall stand for the difference between the upper and lower Darboux sum for f corresponding to $\mathcal{P}$. For $\alpha > 0$, we set

$$S(f;\alpha) = \sup_{n(\mathcal{P})\leqslant\alpha} s(f;\mathcal{P}) .$$

Then the following estimate can be shown.

THEOREM 2.4. (Hlawka [30]). If f is Riemann integrable on E and $x_1,\ldots,x_N$ are points in $[0,1)$ with discrepancy D_N, then

$$\left|\frac{1}{N}\sum_{n=1}^{N} f(x_n) - \int_0^1 f(t)\,dt\right| \leq 3S(f;[D_N^{-1}]^{-1}) .$$

3. The Sequences $(\{n\alpha\})$.

All the error estimates in the preceding section suggest that, in order to arrive at an efficient numerical integration method, we should choose a sequence $x_1,\ldots x_N$ of nodes in E with small discrepancy D_N. For fixed $N \geq 1$, the discrepancy D_N is minimal for the nodes $1/2N$, $3/2N,\ldots,(2N-1)/2N$ (see the remarks following Theorem 1.1). However, if we increase the value of N to N_1, say, in order to obtain higher accuracy, then the minimal sequence $1/2N_1$, $3/2N_1,\ldots,(2N_1-1)/2N_1$ for N_1 may have no single term in common with the minimal sequence for N. Consequently, the results of the previous computations cannot be used. It will therefore be advantageous to have instead available an infinite sequence $\omega = (x_n)$ with uniformly small discrepancies $D_N(\omega)$. For a given $N \geq 1$, we use then the first N terms of ω as the nodes in the numerical integration. If we now increase N to N_1, all the information gathered previously is employed again in the evaluation of

$$\frac{1}{N_1}\sum_{n=1}^{N_1} f(x_n) ,$$

where f is the integrand under consideration.

It is exactly at this point where the theory of diophantine approximations enters the scene. Let α be an irrational number, and consider the infinite sequence $(\{n\alpha\})$, $n = 1,2,\ldots$, where $\{t\} = t-[t]$ denotes the fractional part of the real number t. It turns out that for α's of small type these sequences are very evenly distributed in E, i.e., they have uniformly low discrepancies D_N.

Let us now describe the exact relation between the type of α and the discrepancy of $(\{n\alpha\})$. A systematic study of this relationship was undertaken by Behnke [6], and is based on earlier results of Hardy and Littlewood [21], [22], Hecke [24], Ostrowski [49], and Behnke [5]. We shall give proofs of these results which are much shorter than Behnke's. They will rely on the Erdős-Turán inequality (Theorem 1.4) and very simple estimates for sums of the form

$$\sum_{h=1}^{m} \frac{1}{h\|h\alpha\|} .$$

Here, and in the sequel, $\|t\|$ stands for

$$\|t\| = \min_{n\in Z} |t-n| = \min(\{t\}, 1-\{t\}) .$$

Definition 3.1. Let Ψ be a non-decreasing positive function defined at least for all positive integers. The irrational α is said to be of type $< \Psi$ if

$$q\Psi(q)\|q\alpha\| \geq 1 \quad \text{for all positive integers } q.$$

Definition 3.2. For $\eta \geq 1$, the irrational number α is said to be of type η if η is the infimum of all real numbers $\tau \geq 1$ for which there exists a positive constant $c = c(\tau,\alpha)$ such that α is of type $< \Psi$, where $\Psi(q) = cq^{\tau-1}$.

LEMMA 3.1. The discrepancy $D_N(\omega)$ of $\omega = (\{n\alpha\})$, α irrational, satisfies

$$D_N(\omega) \leq C\left(\frac{1}{m} + \frac{1}{N}\sum_{h=1}^{m} \frac{1}{h\|h\alpha\|}\right)$$

for every positive integer m, where $C > 0$ is an absolute constant.

Proof. For a positive integer h, we have

$$\left|\sum_{n=1}^{N} e^{2\pi ihn\alpha}\right| \leq \frac{2}{|e^{2\pi ih\alpha}-1|} = \frac{1}{|\sin \pi h\alpha|} .$$

But $|\sin \pi h\alpha| = \sin \pi\|h\alpha\|$, and $\sin \pi x \geq 2x$ for $0 \leq x \leq \frac{1}{2}$. Therefore

$$\left|\sum_{n=1}^{N} e^{2\pi ihn\alpha}\right| \leq \frac{1}{2\|h\alpha\|} .$$

The rest follows from Theorem 1.4.

LEMMA 3.2. Let α be of type $< \Psi$. Then

$$\sum_{h=1}^{m} \frac{1}{h\|h\alpha\|} = O(\Psi(2m)\log m + \sum_{h=1}^{m} \frac{\Psi(2h)\log h}{h}) . \tag{3.1}$$

Proof. By Abel's summation formula, we have

$$\sum_{h=1}^{m} \frac{1}{h\|h\alpha\|} = \sum_{h=1}^{m} \frac{s_h}{h(h+1)} + \frac{s_m}{m+1} , \tag{3.2}$$

where $s_h = \sum_{j=1}^{h} \|j\alpha\|^{-1}$. For $0 \leq p < q \leq h$, we obtain

$$\|q\alpha \pm p\alpha\| = \|(q\pm p)\alpha\| \geq \frac{1}{(q\pm p)\Psi(q\pm p)}$$

$$\geq \frac{1}{2h\Psi(2h)} .$$

Using simple properties of the function $\|\cdot\|$, it follows that

$$\left| \|q\alpha\| - \|p\alpha\| \right| \geq \frac{1}{2h\Psi(2h)} = d \quad \text{for} \quad 0 \leq p < q \leq h . \tag{3.3}$$

But (3.3) implies that in each of the intervals $[0,d), [d,2d),\ldots,[hd, (h+1)d)$ there is at most one number of the form $\|j\alpha\|$, $1 \leq j \leq h$, with no such number lying in the first interval. Therefore

$$(3.4) \quad s_h = \sum_{j=1}^{h} \frac{1}{\|j\alpha\|} \leq \sum_{j=1}^{h} \frac{1}{jd} = O(h\Psi(2h)\log h).$$

Together with (3.2) we arrive at the desired estimate.

The estimate in Lemma 3.2 could be improved at the cost of considerable complications by using continued fractions. One would thereby obtain

$$\sum_{h=1}^{m} \frac{1}{h\|h\alpha\|} = O(\log^2 m + \Psi(m) + \sum_{h=1}^{m} \frac{\Psi(h)}{h}) .$$

However, our simple result (3.1) suffices to deduce the desired discrepancy estimates.

<u>THEOREM 3.1</u> Let α be of type η. Then the discrepancy $D_N(\omega)$ of $\omega = (\{n\alpha\})$ satisfies

$$(3.5) \quad D_N(\omega) = O(N^{-\frac{1}{\eta}+\varepsilon}) \quad \text{for every } \varepsilon > 0 .$$

<u>Proof</u>. By Definition 3.2, there will exist for every $\varepsilon > 0$ a constant $c = c(\alpha,\varepsilon) > 0$ such that α is of type $< \Psi$, where $\Psi(q) = cq^{\eta-1+\varepsilon/2}$. It follows then from Lemma 3.2 that

$$\sum_{h=1}^{m} \frac{1}{h\|h\alpha\|} = O(m^{\eta-1+\varepsilon}) .$$

Together with Lemma 3.1 we obtain

$$D_N(\omega) \leqslant C_1\left(\frac{1}{m} + \frac{1}{N} m^{\eta-1+\varepsilon}\right)$$

for every positive integer m, where $C_1 > 0$ only depends on α and ε. Now choose $m = [N^{1/\eta}]$, and we get

$$D_N(\omega) = O(N^{-\frac{1}{\eta}+\varepsilon}) .$$

The estimate (3.5) is best possible in the sense that we also have $D_N(\omega) = \Omega(N^{-1/\eta-\varepsilon})$ for every $\varepsilon > 0$ (Behnke [6]). If α is of constant type, i.e., if α is of type $< \Psi$ where Ψ is a constant function, then the above method yields $ND_N(\omega) = O(\log^2 N)$ for the sequence $\omega = (\{n\alpha\})$. This estimate can be improved by using continued fractions. It is well known that α is of constant type if and only if α has bounded partial quotients in its continued fraction expansion. The subsequent estimate improves earlier results of Ostrowski [49], Behnke [6], and Zaremba [65].

THEOREM 3.2. Let $\alpha = [a_0, a_1, a_2, \ldots]$ be the continued fraction expansion of the irrational α. Suppose that $a_i \leqslant K$ for $i \geqslant 1$. Then the discrepancy $D_N(\omega)$ of $\omega = (\{n\alpha\})$ satisfies

$$ND_N(\omega) \leqslant 3 + \left(\frac{1}{\log \xi} + \frac{K}{\log (K+1)}\right) \log N ,$$

where $\xi = \frac{1+\sqrt{5}}{2}$.

Proof. Let $1 = q_0 \leqslant q_1 < q_2 < \dots$ be the denominators of the convergents to α. For a given $N \geq 1$, there exists $r \geqslant 0$ such that $q_r \leqslant N < q_{r+1}$. By the division algorithm, we have $N = b_r q_r + N_{r-1}$ with $0 \leqslant N_{r-1} < q_r$. We note that $(a_{r+1}+1)q_r \geqslant q_{r+1} > N$, and so $b_r \leqslant a_{r+1}$. If $r > 0$, we may write $N_{r-1} = b_{r-1}q_{r-1} + N_{r-2}$ with $0 \leqslant N_{r-2} < q_{r-1}$. Again we find $b_{r-1} \leqslant a_r$. Continuing in this manner, we arrive at a representation for N of the form

$$N = \sum_{i=0}^{r} b_i q_i$$

with $0 \leqslant b_i \leqslant a_{i+1}$ for $0 \leqslant i \leqslant r$, and $b_r \geqslant 1$.

We decompose the given sequence $\{\alpha\}, \{2\alpha\}, \dots, \{N\alpha\}$ into b_r sequences $(\{n\alpha\})$ where n runs through q_r consecutive integers, into b_{r-1} sequences $(\{n\alpha\})$ where n runs through q_{r-1} consecutive integers, and so on. We estimate first the discrepancy of such a finite sequence $(\{n\alpha\})$ where n runs through q_i consecutive integers, say $n = n_0 + j$ with $1 \leqslant j \leqslant q_i$. If p_i/q_i is the i-th convergent to α, then

$$\alpha = \frac{p_i}{q_i} + \frac{\vartheta}{q_i q_{i+1}} \quad \text{with} \quad |\vartheta| < 1.$$

Therefore

$$\{n\alpha\} = \{n_0\alpha + \frac{jp_i}{q_i} + \frac{j\vartheta}{q_i q_{i+1}}\} .$$

Since g.c.d. $(p_i, q_i) = 1$, the numbers $n_0\alpha + jp_i/q_i$, $1 \leq j \leq q_i$, considered modulo one, form a sequence of q_i equidistant points with distance $1/q_i$. Since

$$\left|\frac{j\vartheta}{q_i q_{i+1}}\right| < \frac{1}{q_{i+1}} \quad \text{for} \quad 1 \leq j \leq q_i ,$$

the sequence $\{n\alpha\}$, $n_0+1 \leq n \leq n_0+q_i$, is obtained by shifting modulo one the elements $\{n_0\alpha + jp_i/q_i\}$, $1 \leq j \leq q_i$, either all to the right or all to the left by a distance less than $1/q_{i+1}$ (the direction of the shift will depend on the sign of ϑ). Therefore, if we order the elements $\{n\alpha\}$, $n_0 + 1 \leq n \leq n_0 + q_i$, according to their magnitude, then the k-th element x_k will satisfy

$$(3.6) \qquad \left|x_k - \frac{2k-1}{2q_i}\right| \leq \frac{1}{2q_i} + \frac{1}{q_{i+1}} .$$

It follows then from Theorem 1.1 that for the discrepancy D_{q_i} of the finite sequence $(\{n\alpha\})$, $n_0+1 \leq n \leq n_0+q_i$, we have

$$D_{q_i} \leq \frac{1}{q_i} + \frac{1}{q_{i+1}} .$$

By [41, Lemma 3.1] and the way in which we decomposed the original sequence $\{\alpha\}, \{2\alpha\}, \ldots, \{N\alpha\}$, we infer from the above that

$$N\, D_N(\omega) \leqslant \sum_{i=0}^{r} b_i \left(1 + \frac{q_i}{q_{i+1}}\right) .$$

Using $b_i \leqslant a_{i+1}$ and $q_{i+1} = a_{i+1}\, q_i + q_{i-1}$, we conclude that

$$(3.7) \qquad N\, D_N(\omega) \leqslant \sum_{i=0}^{r} b_i + \sum_{i=0}^{r} a_{i+1} \frac{q_i}{q_{i+1}}$$

$$\leqslant \sum_{i=0}^{r} b_i + r + 1.$$

We shall now prove that

$$(3.8) \qquad \sum_{i=0}^{r} b_i \leqslant 1 + \frac{K}{\log(K+1)} \log N .$$

We proceed by induction on r. For $N \geqslant 1$, we put $\sigma(N) = \sum_{i=0}^{r} b_i$. If $q_o < q_1$, then the smallest possible r is $r = 0$, and a corresponding N satisfies $1 \leqslant \sigma(N) = N < q_1 \leqslant K$. If $q_o = q_1 = 1$, then the smallest possible r is $r = 1$, and a corresponding N satisfies $1 \leqslant \sigma(N) = N < q_2 \leqslant K + 1$. For the first step in the induction, it therefore suffices to show

$$N \leqslant 1 + \frac{K}{\log(K+1)} \log N \quad \text{for} \quad 1 \leqslant N < K+1.$$

This follows by considering the function $f(x) = x - \frac{K}{\log(K+1)} \log x$ on the interval $1 \leqslant x \leqslant K+1$, and noting that $f(1) = f(K+1) = 1$, and that f is concave upward on the entire interval.

Now take an arbitrary N with $1 < q_r \leq N < q_{r+1}$, and with $N = b_r q_r + N_{r-1}$ with $0 \leq N_{r-1} < q_r$. Suppose for the moment that $N_{r-1} > 0$. Then $\sigma(N) = b_r + \sigma(N_{r-1})$, and the induction hypothesis yields $\sigma(N) \leq 1 + b_r + \frac{K}{\log(K+1)} \log N_{r-1}$. But

$$N > b_r N_{r-1} + N_{r-1} = (b_r+1)N_{r-1},$$

and so

$$\sigma(N) \leq 1 + b_r + \frac{K}{\log(K+1)} \log \frac{N}{b_r+1} .$$

This inequality holds for the case $N_{r-1} = 0$ as well, since then $\sigma(N) = b_r$ and $\frac{N}{b_r+1} = \frac{b_r q_r}{b_r+1} \geq 1$. It remains to prove that $b_r \leq \frac{K}{\log(K+1)} \log(b_r+1)$. But this follows immediately from $1 \leq b_r \leq a_{r+1} \leq K$ and the fact that $g(x) = \frac{x}{\log(x+1)}$ is monotonically increasing for $x > 0$. Therefore (3.8) is established.

To complete the proof, we have to estimate $r + 1$ in terms of N. It follows easily by induction that $q_i \geq \xi^{i-1}$ for $i \geq 0$. Therefore $N \geq q_r \geq \xi^{r-1}$, and so

$$r + 1 \leq 2 + \frac{\log N}{\log \xi} . \tag{3.9}$$

Combining (3.7), (3.8), and (3.9), we arrive at the desired inequality.

The estimate in Theorem 3.2 is best possible as to order of magnitude (compare with the result of Schmidt [55] quoted in section 1). For the sequences $\omega = (\{n\alpha\})$, α irrational, it was already shown by Behnke [6] that $ND_N(\omega) = \Omega(\log N)$.

For $\alpha = \frac{\sqrt{5}\pm 1}{2}$, we can take $K = 1$ in Theorem 3.2. With these values of α, we may therefore expect extremely small discrepancies. A more detailed investigation carried out by Zaremba [65] led to the estimate $ND_N(\omega) \leq \frac{7}{6} \log(6N)$ in this special case. This was improved recently by Sós (unpublished) who showed that $ND_N(\omega) \leq \log N$ for $N \geq 2$.

Progress was also made recently in the estimation of the discrepancy of another type of sequence, namely the van der Corput sequence. Let $n = \sum_{i=0}^{\infty} a_i 2^i$ be the binary expansion of the non-negative integer n. Then define $x_n = \sum_{i=0}^{\infty} a_i 2^{-i-1}$. The infinite sequence $\omega = (x_n)$, $n = 0,1,\ldots,$ is called the van der Corput sequence. Tijdeman (unpublished), improving results of Haber [14], proved that $ND_N(\omega) \leq 1 + \frac{\log N}{3 \log 2}$ and

$$\overline{\lim_{N\to\infty}} \left(ND_N(\omega) - \frac{\log N}{3 \log 2}\right) \geq \frac{4}{9} + \frac{\log 3}{3 \log 2} .$$

For the L^2 discrepancy of the sequence $\omega = (\{n\alpha\})$, we may also give estimates which are essentially best possible. It turns out that the order of magnitude of the L^2 discrepancy of these sequences is practically the same as that of the extreme discrepancy.

THEOREM 3.3. Let α be of type η. Then the L^2 discrepancy $D_N^{(2)}(\omega)$ of $\omega = (\{n\alpha\})$ satisfies $D_N^{(2)}(\omega) = O(N^{-1/\eta+\varepsilon})$ for every $\varepsilon > 0$, but also $D_N^{(2)}(\omega) = \Omega(N^{-1/\eta-\varepsilon})$ for every $\varepsilon > 0$.

Proof. The first assertion follows immediately from Theorem 3.1 and Corollary 1.2. By a result of Behnke [6], we have

$$\frac{1}{N}\left|\sum_{n=1}^{N}(\{n\alpha\} - \tfrac{1}{2})\right| = \Omega(N^{-1/\eta-\varepsilon})$$

for every $\varepsilon > 0$. The second assertion follows then from Theorem 1.3.

THEOREM 3.4. Let $\alpha = [a_0, a_1, a_2, \ldots]$ be the continued fraction expansion of the irrational α. Suppose that $a_i \leq K$ for $i \geq 1$. Then the L^2 discrepancy $D_N^{(2)}(\omega)$ of $\omega = (\{n\alpha\})$ satisfies

$$(ND_N^{(2)}(\omega))^2 \leq (1 + \frac{K}{\log(K+1)} \log N)$$

$$(\frac{13}{3} + \frac{2}{\log \xi} \log N + \frac{K}{3 \log(K+1)} \log N),$$

where $\xi = \frac{1+\sqrt{5}}{2}$.

Proof. We have

$$(ND_N^{(2)}(\omega))^2 = \int_0^1 (A(t;N) - Nt)^2 dt.$$

We decompose the sequence $\{\alpha\}, \{2\alpha\},\ldots,\{N\alpha\}$ into $L = \sum_{i=0}^{r} b_i$ finite subsequences in exactly the same way as in the proof of Theorem 3.2. For $1 \leq m \leq L$, suppose that the m-th subsequence τ_m has N_m elements, and let $A^{(m)}(t;N_m)$ be the counting function with respect to this subsequence. Then

$$A(t;N)-Nt = \sum_{m=1}^{L} (A^{(m)}(t;N_m)-N_m t) \quad \text{for } 0 \leq t \leq 1 .$$

By the Cauchy-Schwarz inequality, we therefore get

$$\int_0^1 (A(t;N)-Nt)^2 dt \leq L \sum_{m=1}^{L} \int_0^1 (A^{(m)}(t;N_m)-N_m t)^2 dt$$

$$= L \sum_{m=1}^{L} (N_m D_{N_m}^{(2)}(\tau_m))^2 .$$

The structure of the sequences τ_m was described in detail in the proof of Theorem 3.2. We note here that $N_m = q_i$ for some i with $0 \leq i \leq r$, where i will of course depend on m. Also, if we order the elements of τ_m according to their magnitude, then by (3.6) the k-th element x_k will satisfy the inequality

$$\left|x_k - \frac{2k-1}{q_i}\right| \leq \frac{1}{2q_i} + \frac{1}{q_{i+1}} .$$

It follows then from Corollary 1.1 that

$$(N_m D_{N_m}^{(2)}(\tau_m))^2 \leqslant \left(\frac{1}{2} + \frac{q_i}{q_{i+1}}\right)^2 + \frac{1}{12}$$

$$= \frac{1}{3} + \frac{q_i}{q_{i+1}} + \frac{q_i^2}{q_{i+1}^2} .$$

Consequently, we get

$$(ND_N^{(2)}(\omega))^2 \leqslant L \sum_{i=0}^{r} b_i \left(\frac{1}{3} + \frac{q_i}{q_{i+1}} + \frac{q_i^2}{q_{i+1}^2}\right) ,$$

since there will be exactly b_i sequences of length q_i among the τ_m, for each i with $0 \leqslant i \leqslant r$. Using the arguments which led to (3.7), we infer that

$$(ND_N^{(2)}(\omega))^2 \leqslant L\left(\frac{L}{3} + 2r + 2\right) .$$

Applying the inequalities (3.8) and (3.9) completes the proof.

We note that we also have $ND_N^{(2)}(\omega) = \Omega(\log N)$, because of

$$\left|\sum_{n=1}^{N} (\{n\alpha\} - \tfrac{1}{2})\right| = \Omega(\log N)$$

(see Ostrowski [49]) and Theorem 1.3.

4. Discrepancy (multidimensional case).

The method of low discrepancy sequences can be extended to any number of dimensions. It is rather obvious how the definition of discrepancy generalizes to the multidimensional case.

For $k \geq 2$, let $E^k = \{(t_1,\ldots,t_k) \in \mathbb{R}^k : 0 \leq t_i \leq 1$ for $1 \leq i \leq k\}$ be the closed k-dimensional unit cube. Let $\underline{x}_1,\ldots,\underline{x}_N$ be a finite sequence of points in E^k with

$$\underline{x}_n = (x_{n1},\ldots,x_{nk}) \quad \text{for} \quad 1 \leq n \leq N.$$

We introduce the remainder function

$$R_N(t_1,\ldots,t_k) = \frac{A(t_1,\ldots,t_k;N)}{N} - t_1 \ldots t_k \quad \text{for} \quad (t_1,\ldots,t_k) \in E^k,$$

where $A(t_1,\ldots,t_k;N)$ denotes the number of elements $\underline{x}_n$, $1 \leq n \leq N$, for which $x_{ni} < t_i$ for $1 \leq i \leq k$.

Definition 4.1. The (extreme) discrepancy D_N of the sequence $\underline{x}_1,\ldots,\underline{x}_N$ in E^k is defined to be

$$D_N = \sup_{(t_1,\ldots,t_k) \in E^k} |R_N(t_1,\ldots,t_k)|.$$

Definition 4.2. The L^2 discrepancy $D_N^{(2)}$ of the sequence $\underline{x}_1,\ldots,\underline{x}_N$ in E^k is defined to be

$$D_N^{(2)} = \left(\int_0^1 \cdots \int_0^1 R_N^2(t_1,\ldots,t_k)\,dt_1\ldots dt_k\right)^{1/2}.$$

For infinite sequences, we adopt conventions analogous to Definition 1.3. In the multidimensional case, the L^2 discrepancy seems to be easier to handle than the extreme discrepancy. A multidimensional analogue to Theorem 1.1 is not yet known, but see Niederreiter [42], [43] for results in this direction. However, Theorem 1.2 can be generalized easily by using the same method of proof.

<u>THEOREM 4.1.</u> (Warnock [61]). The L^2 discrepancy $D_N^{(2)}$ of the sequence $\underline{x}_1,\ldots,\underline{x}_N$ in E^k is given by

$$(D_N^{(2)})^2 = 3^{-k} - \frac{2^{-k+1}}{N} \sum_{n=1}^{N} \prod_{i=1}^{k} (1-x_{ni}^2)$$

$$+ \frac{1}{N^2} \sum_{n=1}^{N} \sum_{m=1}^{N} \prod_{i=1}^{k} (1-\max(x_{ni},x_{mi})) .$$

Lower bounds are known for these discrepancies. The strongest results are available for the case k=2. Here Schmidt [55] has shown $D_N \geq (700N)^{-1} \log N$, and this can even be improved to $D_N \geq (200\,N)^{-1}\log N$ (see Kuipers and Niederreiter [37, Ch. 2]). This result is best possible, apart from the value of the constant (Halton [16]). Again for $k = 2$, we have $D_N^{(2)} \geq cN^{-1}\sqrt{\log N}$ with some absolute constant $c > 0$ (Roth [53]), and this is best possible in the same sense as above (Davenport [9], Halton and Zaremba [19]). For $k > 2$, we have $D_N \geq D_N^{(2)} \geq C_k N^{-1}(\log N)^{\frac{k-1}{2}}$ (Roth [53]), and also $D_N(\omega) = \Omega(N^{-1}(\log N)^{k/2})$ (Kuipers and Niederreiter [37, ch. 2]). One can even show that $D_N^{(2)}(\omega) = \Omega(N^{-1}(\log N)^{k/2})$ (see [46]). It is not known whether these results are best possible.

THEOREM 4.2. For any N points in E^k we have

$$c_k D_N^{\frac{k+2}{2}} \leq D_N^{(2)} \leq D_N$$

with a constant $c_k > 0$ only depending on k.

Proof. The second inequality is obvious. The first inequality follows from a result of Yurinskii [64], by choosing as the two distribution functions the uniform distribution in E^k and the empirical distribution function of the N given points in E^k.

The important Theorem 1.4 can also be extended to several dimensions. For a lattice point $\underline{h} = (h_1,\ldots,h_k)$ in $\mathbb{R}^k$, we define $R(\underline{h}) = \prod_{i=1}^{k} \max(1, |h_i|)$ and $p(\underline{h}) = \max_{1 \leq i \leq k} |h_i|$. We use $\langle .,. \rangle$ to denote the standard inner product in $\mathbb{R}^k$.

THEOREM 4.3. For any N points $\underline{x}_1, \ldots, \underline{x}_N$ in E^k and any positive integer m, the discrepancy D_N of $\underline{x}_1, \ldots, \underline{x}_N$ satisfies

$$D_N \leq 2k^2\left(3^{k+1} + \frac{1}{3}\right)\frac{1}{m} + 2k\left(\frac{k}{\pi}\, 3^k + \frac{1}{5}\right)$$

$$\sum_{0<p(\underline{h})\leq m} R^{-1}(\underline{h}) \left|\frac{1}{N}\sum_{n=1}^{N} e^{2\pi i\langle \underline{h}, \underline{x}_n\rangle}\right|.$$

This theorem, with unspecified constants, was established at about the same time by Koksma [34] and Szüsz.* For the inequality with the specific constants, see Kuipers and Niederreiter [37, Ch. 2].

5. Error Estimates (multidimensional case).

Most of the estimates for the integration error presented in section 2 can be extended to several dimensions. In order to generalize Koksma's inequality, we first have to introduce a suitable concept of total variation.

Definition 5.1. Let f be a function on E^k. For $1 \leq p \leq k$ and $1 \leq i_1 < i_2 < \ldots < i_p \leq k$, we denote by $V^{(p)}(f; i_1, \ldots, i_p)$ the p-dimensional variation in the sense of Vitali of the restriction of f to $E^k_{i_1 \ldots i_p} = \{(t_1, \ldots, t_k) \in E^k : t_j = 1$ whenever j is none of the i_r, $1 \leq r \leq p\}$.

*C.R. Premier Congrès Hongrois, pp. 461-472, Budapest, 1952.

If all the variations $V^{(p)}(f;i_1,\ldots,i_p)$ are finite, the function f is said to be of bounded variation on E^k in the sense of Hardy and Krause.

THEOREM 5.1. (Hlawka [26], Zaremba [66]). Let f be of bounded variation on E^k in the sense of Hardy and Krause, and let ω be a finite sequence of points $\underline{x}_1,\ldots,\underline{x}_N$ in E^k. Then we have

$$\left|\frac{1}{N}\sum_{n=1}^{N} f(\underline{x}_n) - \int_{E^k} f(\underline{t})d\underline{t}\right|$$

$$\leqslant \sum_{p=1}^{k} \sum_{1\leqslant i_1<i_2<\ldots<i_p\leqslant k} V^{(p)}(f;i_1,\ldots,i_p)$$

$$D_N(\omega_{i_1\ldots i_p}) ,$$

where $D_N(\omega_{i_1\ldots i_p})$ is the discrepancy in $E^k_{i_1\ldots i_p}$ of the sequence $\omega_{i_1\ldots i_p}$ obtained by projecting ω onto $E^k_{i_1\ldots i_p}$.

In crucial steps, the proof of the above theorem rests on a generalized Abel summation formula. Zaremba [66] elucidated this principle by establishing a formula for multidimensional integration by parts from which the above inequality can be deduced rather easily (compare with the proof of Lemma 2.1).

THEOREM 5.2. (Sobol' [56], Zaremba [66]). Let $f(\underline{t}) = f(t_1,\dots,t_k)$ be a function on E^k for which the partial derivative $\frac{\partial^k f}{\partial t_1 \dots \partial t_k}$ is continuous on E^k, and let ω be a finite sequence of points $\underline{x}_1,\dots,\underline{x}_N$ in E^k. Then

$$\left|\frac{1}{N}\sum_{n=1}^{N} f(\underline{x}_n) - \int_{E^k} f(\underline{t})d\underline{t}\right|$$

$$\leqslant \sum_{p=1}^{k} \sum_{1\leqslant i_1<i_2<\dots<i_p\leqslant k} D_N^{(2)}(\omega_{i_1\dots i_p})$$

$$\|f_{t_{i_1}\dots t_{i_p}}\|_2 ,$$

where $D_N^{(2)}(\omega_{i_1\dots i_p})$ is the L^2 discrepancy in $E^k_{i_1\dots i_p}$ of the sequence $\omega_{i_1\dots i_p}$ defined as in Theorem 5.1, and $\|f_{t_{i_1}\dots t_{i_p}}\|_2$ is the L^2 norm on $E^k_{i_1\dots i_p}$ of the restriction of the partial derivative $\frac{\partial^p f}{\partial t_{i_1}\dots\partial t_{i_p}}$ to $E^k_{i_1\dots i_p}$.

The proof of Theorem 5.2 proceeds along the same lines as that of Theorem 5.1, but at a certain point one applies the Cauchy-Schwarz inequality rather than the triangle inequality (compare with the proof of Theorem 2.2).

For a bounded function f on E^k and a partition $\mathcal{P}$ of $U^k = \{(t_1,\ldots,t_k) \in \mathbb{R}^k : 0 \leq t_j < 1 \text{ for } 1 \leq j \leq k\}$ into half-open intervals, we denote by $s(f;\mathcal{P})$ the difference between the upper and lower Darboux sum for f corresponding to the partition $\mathcal{P}$. The norm $n(\mathcal{P})$ of the partition $\mathcal{P}$ is defined as the maximum length of all the edges of the intervals in $\mathcal{P}$. For a real number $\alpha > 0$, we set

$$S(f;\alpha) = \sup_{n(\mathcal{P}) \leq \alpha} s(f;\mathcal{P}) .$$

The following error estimates hold.

<u>THEOREM 5.3</u>. (Hlawka [30]). For a Rieman integrable function f on E^k and points $\underline{x}_1,\ldots,\underline{x}_N$ in U^k with discrepancy D_N, we have

$$\left|\frac{1}{N}\sum_{n=1}^{N} f(\underline{x}_n) - \int_{E^k} f(\underline{t})d\underline{t}\right|$$

$$\leq (2^{2k-1} + 1)S(f;[D_N^{-1}]^{-\frac{1}{k}}).$$

<u>COROLLARY 5.1</u>. (Hlawka [30]). If f is continuous on E^k with modulus of continuity M, and if the points $\underline{x}_1,\ldots,\underline{x}_N$ in U^k have discrepancy D_N, then

$$\left|\frac{1}{N}\sum_{n=1}^{N} f(\underline{x}_n) - \int_{E^k} f(\underline{t})d\underline{t}\right|$$
$$\leq (2^{2k-1} + 1)M([D_N^{-1}]^{-\frac{1}{k}}).$$

Theorem 5.3 and Corollary 5.1 are of course extensions of Theorems 2.3 and 2.4 to the multi-dimensional case, with the constant $2^{2k-1} + 1$ probably being too large.

6. Special Sequences.

In the light of the results in the preceding section, we will search for sequences in E^k with small discrepancy. An important class of favorable sequences will be of the form $((\{n\alpha_1\},\ldots,\{n\alpha_k\}))$, $n = 1,2,\ldots$, where $(\alpha_1,\ldots,\alpha_k)$ is a k-tuple of badly approximable irrationals in the sense of simultaneous diophantine approximations. We again introduce a notion of type. For lattice points $\underline{h}$ in $\mathbb{R}^k$, we use $R(\underline{h})$ as defined in section 4.

Definition 6.1. For a real number η, a k-tuple $\underline{\alpha} = (\alpha_1,\ldots,\alpha_k)$ of irrationals is said to be of type η if η is the infimum of all numbers σ for which there exists a positive constant $c = c(\sigma;\alpha_1,\ldots,\alpha_k)$ such that

$$(6.1) \qquad R^{\sigma}(\underline{h})\ \|\langle\underline{h},\underline{\alpha}\rangle\| \geq c$$

holds for all lattice points $\underline{h} \neq \underline{0}$ in $\mathbb{R}^k$, where $\langle .,. \rangle$ denotes the standard inner product in $\mathbb{R}^k$.

It follows easily from the Minkowski linear forms theorem that we will always have $\eta \geq 1$. It should be remarked that for certain $\underline{\alpha}$ the inequality (6.1) need not hold for any finite σ. Such an $\underline{\alpha}$ could be called of infinite type. On the other hand, k-tuples $\underline{\alpha}$ with type $\eta = 1$ have recently been established. Schmidt [54] has shown (thereby generalizing Roth's theorem) that $\underline{\alpha} = (\alpha_1, \ldots, \alpha_k)$, with real algebraic numbers α_i for which $1, \alpha_1, \ldots, \alpha_k$ are linearly independent over the rationals, is of type $\eta = 1$. It follows from a result of Baker [3] that $\underline{\alpha} = (e^{r_1}, \ldots, e^{r_k})$, with distinct nonzero rationals $r_1, \ldots, r_k$, is of type $\eta = 1$ (he shows in fact a considerably stronger statement).

In the multidimensional case, we have again a strong correlation between the type of $\underline{\alpha} = (\alpha_1, \ldots, \alpha_k)$ and the evenness of distribution of the sequence associated with α.

THEOREM 6.1. Let $\underline{\alpha} = (\alpha_1, \ldots, \alpha_k)$ be a k-tuple of irrationals of type $\eta = 1$. Then the discrepancy $D_N(\omega)$ of the sequence $\omega = ((\{n\alpha_1\}, \ldots, \{n\alpha_k\}))$, $n = 1, 2, \ldots$, satisfies

$$D_N(\omega) = O(N^{-1+\varepsilon}) \quad \text{for every} \quad \varepsilon > 0 .$$

Proof. Since we use the same method as in [43, Theorem 9], we just sketch the proof briefly. Using Theorem 4.3, we conclude as in Lemma 3.1 that

$$D_N(\omega) \leqslant C_k\Big(\frac{1}{m} + \frac{1}{N} \sum_{0<p(\underline{h})\leqslant m} R^{-1}(\underline{h})\,\|\langle\underline{h},\underline{\alpha}\rangle\|^{-1}\Big) \tag{6.2}$$

for every positive integer m, where C_k only depends on the dimension k. By iterated Abel summation, we arrive at the following identity:

$$\sum_{0<p(\underline{h})\leqslant m} R^{-1}(\underline{h})\,\|\langle\underline{h},\underline{\alpha}\rangle\|^{-1} = \sum_{n_1,\ldots,n_k=1}^{m} f(n_1,\ldots,n_k) \sum^{*}_{\substack{\underline{h}=(h_1,\ldots,h_k)\\ |h_j|\leqslant n_j}} \|\langle\underline{h},\underline{\alpha}\rangle\|^{-1}, \tag{6.3}$$

where $f(n_1,\ldots,n_k) = \prod_{i=1}^{k} g_m(n_i)$ with $g_m(n) = \frac{1}{n(n+1)}$ for $1 \leqslant n < m$ and $g_m(m) = \frac{1}{m}$, and where the asterisk means deletion of the origin from the range of summation. Using the assumption on the type of $\underline{\alpha}$ and the idea which led to (3.4), one can prove that, for every $\varepsilon > 0$, we have

$$\sum^{*}_{\substack{\underline{h}=(h_1,\ldots,h_k)\\ |h_j|\leqslant n_j}} \|\langle\underline{h},\underline{\alpha}\rangle\|^{-1} \leqslant L(n_1\ldots n_k)^{1+\varepsilon}$$

with a constant L independent of $n_1,\ldots,n_k$. Going back to (6.3) we conclude that

$$\sum_{0<p(\underline{h})\leq m} R^{-1}(\underline{h})\,\|\langle\underline{h},\underline{\alpha}\rangle\|^{-1} \leq L\Big(\sum_{n=1}^{m} g_m(n)n^{1+\varepsilon}\Big)^k .$$

From the definition of g_m, it is easy to show that

$$\sum_{n=1}^{m} g_m(n)n^{1+\varepsilon} = O(m^{\varepsilon}) .$$

Therefore we get

$$\sum_{0<p(\underline{h})\leq m} R^{-1}(\underline{h})\,\|\langle\underline{h},\underline{\alpha}\rangle\|^{-1} = O(m^{k\varepsilon}) .$$

An application of (6.2) yields

$$D_N(\omega) \leq C_k'\Big(\frac{1}{m} + \frac{m^{k\varepsilon}}{N}\Big)$$

with C_k' independent of m and N. Choosing $m = N$ completes the proof.

Using this method, discrepancy estimates for k-tuples $\underline{\alpha}$ of finite type $\eta > 1$ may also be established.

Multidimensional versions of the van der Corput sequence (see section 3) have also been studied. We refer to the detailed survey of Halton [17, section 2.5] and to the papers of Sobol' [56], [57], [58], [59]. The results may be briefly summarized as follows. There exist finite sequences of N points in E^k with $D_N = O(N^{-1}\log^{k-1} N)$, and there are infinite sequences ω in E^k with $D_N(\omega) = O(N^{-1} \log^k N)$. It is conjectured that these orders of magnitude are best possible (compare with the discussion in section 4). As to the L^2 discrepancy, a significant contribution was made by Halton and Zaremba [19] who constructed finite sequences in E^2 of the van der Corput type with $D_N^{(2)} = O(N^{-1}\log^{1/2} N)$. This shows not only that the L^2 discrepancy can be genuinely smaller than the extreme discrepancy, but it also suggests that Roth's lower bound for $D_N^{(2)}$ (see section 4) may be best possible for all $k \geq 2$.

7. Other Integration Domains.

So far, we restricted the discussion to the case of E^k being the integration domain. But the method of low discrepancy sequences can be adapted to work for more general classes of integration domains as well. The first important step in this direction was undertaken by Zaremba [68]. The emphasis in his paper is on convex subsets of E^k as integration domains.

We shall study a much wider class of integration domains. To this end, we need, first of all, some definitions. In this section, M' will always stand for the complement in E^k of a subset M of E^k.

Definition 7.1. Let $d(.,.)$ be the standard Euclidean distance in $\mathbb{R}^k$. For a subset M of E^k and $\varepsilon > 0$, we define

$$M_\varepsilon = \{\underline{x} \in E^k : d(\underline{x},\underline{y}) < \varepsilon \quad \text{for some} \quad \underline{y} \in M\} ,$$

and

$$M_{-\varepsilon} = ((M')_\varepsilon)' = \{\underline{x} \in E^k : d(\underline{x},\underline{y}) \geq \varepsilon \text{ for all } \underline{y} \in M'\} .$$

The set M_ε is a superset of M open in the relative topology in E^k, whereas $M_{-\varepsilon}$ is a closed subset of M. In the sequel, λ_k will denote Lebesgue measure in E^k.

Definition 7.2. Let r be a positive nondecreasing function defined for all positive numbers which satisfies $\lim_{\varepsilon \to 0+} r(\varepsilon) = 0$. Then $\mathcal{M}_r^{(k)}$ is defined as the family of all λ_k-measurable subsets M of E^k for which the inequalities

$$\lambda_k(M_\varepsilon \setminus M) \leq r(\varepsilon) \quad \text{and} \quad \lambda_k(M \setminus M_{-\varepsilon}) \leq r(\varepsilon)$$

hold for all $\varepsilon > 0$.

At this point, let us make the following observations. If ∂M denotes the topological boundary of a subset M of E^k, then clearly $\partial M \subset M_\varepsilon \setminus M_{-\varepsilon}$ for every $\varepsilon > 0$. It follows therefore that every $M \in \mathcal{M}_r^{(k)}$ is a λ_k-continuity set, i.e., $\lambda_k(\partial M) = 0$ for every $M \in \mathcal{M}_r^{(k)}$. In particular, every $M \in \mathcal{M}_r^{(k)}$ is Jordan measurable. On the other hand, the union $\cup \mathcal{M}_r^{(k)}$, taken over all possible r, just comprises all λ_k-continuity sets in E^k, and so all Jordan measurable subsets of E^k.

LEMMA 7.1. Every λ_k-continuity set in E^k belongs to some $M_r^{(k)}$.

Proof. Let M be a given λ_k-continuity set in E^k. Define a function r by $r(\varepsilon) = \max(\lambda_k(M_\varepsilon \setminus M), \lambda_k(M \setminus M_{-\varepsilon})) + \varepsilon$ for $\varepsilon > 0$. Then r is positive and nondecreasing. Once we have shown $\lim_{\varepsilon \to 0+} r(\varepsilon) = 0$, then obviously $M \in M_r^{(k)}$. Now

$$\partial M = \bigcap_{n=1}^{\infty} (M_{1/n} \setminus M_{-1/n}) ,$$

and so $\lim_{n\to\infty} \lambda_k(M_{1/n} \setminus M_{-1/n}) = 0$. This implies immediately that $\lim_{\varepsilon \to 0+} r(\varepsilon) = 0$.

Definition 7.3. Let ω be a finite sequence of points $\underline{x}_1, \ldots, \underline{x}_N$ in E^k. For r as in Definition 7.2, we set

$$D_N(\omega; M_r^{(k)}) = \sup_{M \in M_r^{(k)}} \left| \frac{A(M;N)}{N} - \lambda_k(M) \right| ,$$

where $A(M;N)$ is the number of n, $1 \leq n \leq N$, with $\underline{x}_n \in M$.

Apart from its generality, the class $M_r^{(k)}$ offers two more reasons for its practicality. Firstly, when approximately computing the integral of a reasonably regular function over a set $M \in M_r^{(k)}$, the resulting error can still be estimated in terms of discrepancies. Even more, the error bound will be uniform in all members M of $M_r^{(k)}$ (see Theorem 7.1). Secondly, the discrepancy $D_N(\omega; M_r^{(k)})$ associated with $M_r^{(k)}$ still satisfies

$\lim_{N\to\infty} D_N(\omega;M_r^{(k)}) = 0$ for any infinite sequence ω uniformly distributed in E^k (this will follow from Theorem 7.2), and nontrivial upper bounds for the discrepancy $D_N(\omega;M_r^{(k)})$ of interesting classes of sequences ω can be established on the basis of Theorem 7.2. It should be noted that uniform conditions on the size of the "ε-boundaries" of sets M in a given class M of Borel subsets of E^k are necessary to guarantee that the discrepancy associated with M tends to zero for every uniformly distributed sequence in E^k (see Billingsley and Topsøe [7]). Thus, the present viewpoint is essentially the most general one which can be adopted in the method of low discrepancy sequences.

In the sequel, U^k shall denote the half-open k-dimensional unit cube $U^k = \{(t_1,\ldots,t_k)\in \mathbb{R}^k: 0 \leq t_j < 1 \text{ for } 1 \leq j \leq k\}$.

THEOREM 7.1. Let f be of bounded variation on E^k in the sense of Hardy and Krause, and let ω be a finite sequence of points $\underline{x}_1,\ldots,\underline{x}_N$ in U^k. Then, for every $M\in M_r^{(k)}$ we have

$$
\begin{aligned}
&\left|\frac{1}{N}\sum_{\underline{x}_n\in M} f(\underline{x}_n) - \int_M f d\lambda_k\right| \\
(7.1)\qquad &\leq D_N(\omega;M_s^{(k)})(|f(1,\ldots,1)|
\end{aligned}
$$

$$+ \sum_{p=1}^{k} \sum_{1 \leq i_1 < i_2 < \ldots < i_p \leq k} V^{(p)}(f; i_1, \ldots, i_p)) ,$$

where s is the function $s(\varepsilon) = r(\varepsilon) + 2k\varepsilon$.

<u>Proof.</u> By the remarks following Definition 7.2, every $M \in M_r^{(k)}$ is Jordan measurable. By [68, Proposition 2.1], we get then the inequality

$$\left| \frac{1}{N} \sum_{\underline{x}_n \in M} f(\underline{x}_n) - \int_M f d\lambda_k \right|$$

$$\leq |f(1,\ldots,1) g_M(1,\ldots,1)|$$

$$+ \sum_{p=1}^{k} \sum_{1 \leq i_1 < i_2 < \ldots < i_p \leq k} V^{(p)}(f; i_1, \ldots i_p)$$

$$\sup_{(t_1,\ldots,t_k) \in E^k} |g_M(t_1,\ldots,t_k)| ,$$

where g_M is defined by

$$g_M(t_1,\ldots,t_k) = \frac{1}{N} A(M \cap I_{t_1,\ldots,t_k}; N)$$

$$- \lambda_k(M \cap I_{t_1,\ldots,t_k})$$

for $(t_1,\ldots,t_k) \in E^k$, with $I_{t_1,\ldots,t_k}$ being the interval $I_{t_1,\ldots,t_k} = [0,t_1) \times \ldots \times [0,t_k)$.

(Note: we have corrected some minor typographical errors in the original statement of this theorem).

To complete the proof, it suffices then to show that $M \in \mathcal{M}_r^{(k)}$ implies that $M \cap I_{t_1,\ldots,t_k} \in \mathcal{M}_s^{(k)}$ for all $(t_1,\ldots,t_k) \in E^k$. But, for any two subsets A, B of E^k and any $\varepsilon > 0$ we have $(A \cap B)_\varepsilon \setminus (A \cap B) \subset (A_\varepsilon \setminus A) \cup (B_\varepsilon \setminus B)$. Therefore

$$\lambda_k((M \cap I_{t_1,\ldots,t_k})_\varepsilon \setminus (M \cap I_{t_1,\ldots,t_k}))$$

$$\leqslant \lambda_k(M_\varepsilon \setminus M) + \lambda_k((I_{t_1,\ldots,t_k})_\varepsilon \setminus I_{t_1,\ldots,t_k})$$

$$\leqslant r(\varepsilon) + k\varepsilon .$$

A similar argument shows

$$\lambda_k((M \cap I_{t_1,\ldots,t_k}) \setminus (M \cap I_{t_1,\ldots,t_k})_{-\varepsilon})$$

$$\leqslant \lambda_k(M \setminus M_{-\varepsilon}) + \lambda_k(I_{t_1,\ldots,t_k} \setminus (I_{t_1,\ldots,t_k})_{-\varepsilon})$$

$$\leqslant r(\varepsilon) + 2k\varepsilon.$$

It may be possible to replace the function s in (7.1) by the function r itself. If f is only Riemann integrable on E^k, one may still estimate the integration error

$$\left|\frac{1}{N} \sum_{\underline{x}_n \in M} f(\underline{x}_n) - \int_M f d\lambda_k\right|$$

by simply using Theorem 5.3 with the Riemann integrable function fc_M, where c_M is the characteristic function of the integration domain $M \in \mathcal{M}_r^{(k)}$.

THEOREM 7.2. The discrepancy $D_N(\omega;\mathcal{M}_r^{(k)})$ and the extreme discrepancy D_N of any sequence ω of N points in U^k are related by

$$(7.2) \qquad D_N(\omega;\mathcal{M}_r^{(k)}) \leq (4^k + 2)r(\sqrt{k}[D_N^{-1}]^{-\frac{1}{k}}) .$$

Proof. For a given $M \in \mathcal{M}_r^{(k)}$, we use Theorem 5.3 with $f = c_M$, and we obtain

$$(7.3) \qquad \left|\frac{A(M;N)}{N} - \lambda_k(M)\right| \leq (2^{2k-1}+ 1)S(c_M;[D_N^{-1}]^{-\frac{1}{k}}).$$

Let $\mathcal{P}$ be a partition of U^k into half-open subintervals. Define $P^{(1)}$ to be the union of all those intervals from $\mathcal{P}$ which are entirely contained in M. Define $P^{(2)}$ to be the union of all those intervals from $\mathcal{P}$ which have a nonvoid intersection with M. We note that, with the notation introduced in section 5, we have then

$$(7.4) \qquad s(c_M;\mathcal{P}) = \lambda_k(P^{(2)}) - \lambda_k(P^{(1)}) .$$

Now suppose that $n(\mathcal{P}) \leq \alpha$, say. Then any two points from one and the same interval in $\mathcal{P}$ have a distance less than $\varepsilon = \alpha\sqrt{k}$. Therefore, each point of $P^{(2)}$ has a distance less than ε from some point of M. In other words, we have $P^{(2)} \subset M_\varepsilon$. In a completely analogous way, one shows that $P^{(1)} \supset M_{-\varepsilon} \cap U^k$ with the same ε as above, and so

$$\lambda_k(P^{(2)}) - \lambda_k(P^{(1)}) \leq \lambda_k(M_\varepsilon) - \lambda_k(M_{-\varepsilon}) .$$

Since $M \in \mathcal{M}_r^{(k)}$, we conclude that

$$(7.5) \qquad \lambda_k(P^{(2)}) - \lambda_k(P^{(1)}) \leq 2r(\alpha\sqrt{k}) .$$

Combining (7.4) and (7.5), and taking the supremum over all $\mathcal{P}$ with $n(\mathcal{P}) \leq \alpha$, we infer that

$$(7.6) \qquad S(c_M;\alpha) \leq 2r(\alpha\sqrt{k}) .$$

The inequalities (7.3) and (7.6) together imply then (7.2).

Let us now restrict our attention to the class $\mathcal{C}$ of convex subsets of E^k. Results about this class of integration domains are implicitly contained in the above considerations, since $\mathcal{C} \subset \mathcal{M}_r^{(k)}$ with $r(\varepsilon) = 2k\varepsilon$ for $\varepsilon > 0$ (see the proof of Lemma 7.2). However, using special properties of convex sets, we can arrive at much stronger theorems. We are led to the notion of isotrope (or isotropic) discrepancy.

<u>Definition 7.4</u>. The isotrope discrepancy J_N of the points $\underline{x}_1,\ldots,\underline{x}_N$ in E^k is given by

$$J_N = \sup_{C \in \mathcal{C}} \left| \frac{A(C;N)}{N} - \lambda_k(C) \right| .$$

The theory of isotrope discrepancy is only in its initial stage (as is the theory of the discrepancies $D_N(\omega; M_r^{(k)})$). The notion was suggested by Hlawka [28], and studied in more detail by Zaremba [68] and Niederreiter [42], [43]. A uniform law of the iterated logarithm for J_N in the case $k = 2$ was established by Philipp [51]. The following results seem to be of interest in the present context.

THEOREM 7.3. (Niederreiter [42]). The isotrope discrepancy J_N of a finite sequence of points in E^k is also given by

$$J_N = \sup_{Q \in q} \left| \frac{A(Q;N)}{N} - \lambda_k(Q) \right| ,$$

where q is the class of all closed or open convex polytopes Q in E^k satisfying the following condition: each face of Q is either lying entirely on the boundary of E^k or contains a point of the sequence.

THEOREM 7.4. (Zaremba [68]). Let C be a convex subset of E^k, let f be of bounded variation on E^k in the sense of Hardy and Krause, and let $\underline{x}_1, \ldots, \underline{x}_N$ be points in U^k with isotrope discrepancy J_N. Then

$$\left| \frac{1}{N} \sum_{\underline{x}_n \in C} f(\underline{x}_n) - \int_C f \, d\lambda_k \right|$$

$$\leqslant J_N(|f(1,\ldots,1)|$$

$$+ \sum_{p=1}^{k} \sum_{1 \leq i_1 < i_2 < \ldots < i_p \leq k} V^{(p)}(f; i_1, \ldots, i_p)) .$$

THEOREM 7.5. (Zaremba [68]). For the isotrope discrepancy J_N of any N points in E^k we have

$$J_N \geq C_k N^{-\frac{k+1}{2k}}$$

with a positive constant C_k only depending on k.

THEOREM 7.6. (Niederreiter [43]). The isotrope discrepancy J_N and the extreme discrepancy D_N of any N points in E^k are related by

$$(7.7) \qquad J_N \leq (8k\sqrt{k} + 2) D_N^{\frac{1}{k}} .$$

Some remarks on Theorem 7.6 are in order. In [43], J_N was estimated in terms of the discrepancy extended over all half-open subintervals of U^k. In the above statement, we used that the latter discrepancy is dominated by $2^k D_N$. Furthermore, the inequality (7.7) was stated in [43] only for the case where the given points are contained in U^k. This offers advantages in the proof (compare with Theorem 7.2). However, by using a simple continuity argument based on the subsequent lemma, the inequality is then easily extended to the slightly more general case considered in Theorem 7.6. The result for D_N analogous to the lemma below is well known (see Niederreiter

[40, Satz 2.4]).

LEMMA 7.2. Let $\omega^{(1)}$ be the finite sequence $\underline{x}_1^{(1)},\ldots,\underline{x}_N^{(1)}$ in E^k with isotrope discrepancy $J_N^{(1)}$, and let $\omega^{(2)}$ be the finite sequence $\underline{x}_1^{(2)},\ldots,\underline{x}_N^{(2)}$ in E^k with isotrope discrepancy $J_N^{(2)}$. Suppose that $d(\underline{x}_n^{(1)}, \underline{x}_n^{(2)}) < \varepsilon$ for $1 \leqslant n \leqslant N$. Then

$$|J_N^{(1)} - J_N^{(2)}| \leqslant 2k\varepsilon.$$

Proof. Let $A^{(1)}$ and $A^{(2)}$ be the counting functions corresponding to $\omega^{(1)}$ and $\omega^{(2)}$ respectively. Let C be a given convex subset of E^k. We note that, whenever $\underline{x}_n^{(2)} \in C$, then $\underline{x}_n^{(1)} \in C_\varepsilon$. Therefore $A^{(2)}(C;N) \leqslant A^{(1)}(C_\varepsilon;N)$. Consequently, we obtain

$$\frac{A^{(2)}(C;N)}{N} - \lambda_k(C) \leqslant \left(\frac{A^{(1)}(C_\varepsilon;N)}{N} - \lambda_k(C_\varepsilon)\right) \tag{7.8}$$
$$+ \lambda_k(C_\varepsilon \setminus C) \leqslant J_N^{(1)} + 2k\varepsilon,$$

where we used that C_ε is again convex, and that $\lambda_k(C_\varepsilon \setminus C) \leqslant 2k\varepsilon$. The latter assertion can be shown as follows.

In the proof of [43, Theorem 11], the inequality $\lambda_k(P_\varepsilon \setminus P) \leqslant 2k\varepsilon$ was established for convex polytopes P in E^k, by arguing that $\lambda_k(P_\varepsilon \setminus P)$ is at most ε times the surface of P_ε, and the surface of P_ε is at most the surface of E^k, which is $2k$. For two nonempty subsets A and B

of E^k, we define their distance by

$$\beta(A,B) = \inf\{\delta > 0: A \subset B_\delta \text{ and } B \subset A_\delta\}.$$

If A and B are convex, then we have a slightly modified Blaschke distance ([8, p. 34]; the only difference is that in the definition of M_ε we admit only points from E^k). Let us further note that $(A_\delta)_\varepsilon = (A_\varepsilon)_\delta$ for any $A \subset E^k$ and any $\delta > 0$ and $\varepsilon > 0$. This is due to the fact that a point $\underline{z} \in (A_\delta)_\varepsilon$, which, by definition, can be reached from some $\underline{x} \in A$ by first going for a distance less than δ to a point $\underline{y} \in E^k$, and then going for a distance less than ε from $\underline{y}$ to $\underline{z}$, can also be reached from $\underline{x}$ by first going for a distance less than ε and then for a distance less then δ, if we simply pass along the line segment joining $\underline{x}$ and $\underline{z}$. Next we claim that

$$(7.9) \qquad \beta(A_\varepsilon, B_\varepsilon) \leq \beta(A,B)$$

for any two nonempty subsets A and B of E^k and any $\varepsilon > 0$. Let $\delta > 0$ be such that $A \subset B_\delta$ and $B \subset A_\delta$. Then $A_\varepsilon \subset (B_\delta)_\varepsilon = (B_\varepsilon)_\delta$ and $B_\varepsilon \subset (A_\delta)_\varepsilon = (A_\varepsilon)_\delta$, and the definition of β already implies (7.9). Finally, let now C be a nonempty convex subset of E^k. We choose a sequence $P^{(m)}$, $m = 1,2,\ldots$, of nonempty convex polytopes converging to C in the (modified) Blaschke distance. Then from (7.9) we infer that $P_\varepsilon^{(m)}$ converges to C_ε in the Blaschke distance. Since the Lebesgue measure λ_k is continuous with respect to the

Blaschke distance ([8, p. 38]), it follows that $\lim_{m\to\infty} \lambda_k(P_\varepsilon^{(m)} \setminus P^{(m)}) = \lambda_k(C_\varepsilon \setminus C)$. Thus $\lambda_k(C_\varepsilon \setminus C) \leqslant 2k\varepsilon$ is now a consequence of

$$\lambda_k(P_\varepsilon^{(m)} \setminus P^{(m)}) \leqslant 2k\varepsilon \quad \text{for all} \quad m \geqslant 1 .$$

Returning to the sequences $\omega^{(1)}$ and $\omega^{(2)}$, we note that, whenever $\underline{x}_n^{(1)} \in C_{-\varepsilon}$, then $\underline{x}_n^{(2)} \in C$. Therefore $A^{(2)}(C;N) \geqslant A^{(1)}(C_{-\varepsilon};N)$, and so

$$\begin{aligned} \frac{A^{(2)}(C;N)}{N} - \lambda_k(C) &\geqslant \Big(\frac{A^{(1)}(C_{-\varepsilon};N)}{N} \\ &- \lambda_k(C_{-\varepsilon})\Big) - \lambda_k(C \setminus C_{-\varepsilon}) \\ &\geqslant - J_N^{(1)} - 2k\varepsilon , \end{aligned} \tag{7.10}$$

since $C_{-\varepsilon} \in \mathcal{C}$, and $\lambda_k(C \setminus C_{-\varepsilon}) \leqslant 2k\varepsilon$ can be verified along the same lines as above. Combining (7.8) and (7.10), we get

$$J_N^{(2)} \leqslant J_N^{(1)} + 2k\varepsilon .$$

By interchanging the roles of $\omega^{(1)}$ and $\omega^{(2)}$, we may deduce the inequality of the lemma.

By invoking Theorem 7.3, we could have carried out the above proof by only working with convex polytopes. However, we wanted to prove in the same run that $\mathcal{C} \subset M_r^{(k)}$ with $r(\varepsilon) = 2k\varepsilon$, a claim we made earlier.

Part B. The Fourier Series Method.

8. Basic Results.

The method to be described now can be traced back to papers of Richtmyer [52] and Peck [50]. For a detailed bibliography on the method, we refer to Niederreiter [44] and Kuipers and Niederreiter [37, section 2.5].

The method applies to integrands f on E^k which are periodic with period 1 in each variable, and which satisfy certain conditions on the order of magnitude of their Fourier coefficients. Later on in this section we shall discuss the possibility of alleviating the very stringent periodicity requirement. The integral $\int_{E_k} f(\underline{t})d\underline{t}$ is again approximated by arithmetic means of function values of f at certain nodes, namely by $\frac{1}{N}\sum_{n=1}^{N} f(n\underline{\alpha})$, where $\underline{\alpha}$ is a suitably chosen point in $\mathbb{R}^k$, and where we have tacitly assumed that f is extended periodically to $\mathbb{R}^k$ (otherwise we have to use the fractional parts of the points $n\underline{\alpha}$). The success of the method depends on a judicious choice of $\underline{\alpha}$, and here again the theory of diophantine approximations will be helpful.

Now to the precise statement of the conditions under which the method operates. Let $f(\underline{t}) = f(t_1,\ldots,t_k)$ be a function on $\mathbb{R}^k$ which is periodic with period 1 in each variable and which can be expanded into a multiple Fourier series

$$f(\underline{t}) = \sum_{\underline{h}} c_{\underline{h}} e^{2\pi i<\underline{h},\underline{t}>} ,$$

where $<.,.>$ denotes again the standard inner product in $\mathbb{R}^k$, and where the sum is extended over all lattice points $\underline{h}$ in $\mathbb{R}^k$. The function f is said to belong to the class $\mathcal{E}^s$ for some $s > 1$, if there exists a constant M (which may depend on f) such that

(8.1) $|c_{\underline{h}}| \leqslant MR^{-s}(\underline{h})$ for all $\underline{h} \neq \underline{0}$.

Here $R(\underline{h})$ has the same meaning as in section 4. We note that in condition (8.1) we may of course as well accept the lattice point $\underline{h} = \underline{0}$, but this might force the constant M to be unnecessarily large. This is undesirable since the value of M enters into explicit estimates for the integration error.

Whenever $f \in \mathcal{E}^s$ for some $s > 1$, the Fourier series for f will be absolutely convergent.This observation will justify various manipulations which have to be carried out in the subsequent proofs. At this point, let us also mention two sufficient conditions which guarantee that $f \in \mathcal{E}^s$, and which, in all practical circumstances, are much easier to verify than (8.1). Let s be an integer greater than unity, and suppose all the partial derivatives

$$\frac{\partial^{q_1+\ldots+q_k} f}{\partial t_1^{q_1} \ldots \partial t_k^{q_k}}$$

with $0 \leq q_j \leq s$ for $1 \leq j \leq k$ exist and are continuous in $\mathbb{R}^k$; then $f \in \mathcal{E}^s$ (Zaremba [66]). Let s again be an integer greater than unity, and suppose all the partial derivatives

$$\frac{\partial^{q_1+\ldots+q_k} f}{\partial t_1^{q_1} \ldots \partial t_k^{q_k}} \quad \text{with} \quad 0 \leq q_j \leq s-1$$

for $1 \leq j \leq k$ exist and are of bounded variation over E^k in the sense of Hardy and Krause; then $f \in \mathcal{E}^s$ (Zaremba [66]). In these cases, a possible value for M in (8.1) may even be given explicitly ([66, Proposition 8]).

For k-tuples $\underline{\alpha} = (\alpha_1, \ldots, \alpha_k)$ of irrationals, we shall use again the notion of type which was introduced in Definition 6.1. The following result provides the theoretical basis for the method. In the proof, we use techniques from [23] and [44].

THEOREM 8.1. Let $\underline{\alpha}$ be a k-tuple of irrationals of type η. Then we have

$$\frac{1}{N} \sum_{n=1}^{N} f(n\underline{\alpha}) - \int_{E^k} f(\underline{t})d\underline{t} = O(N^{-1})$$

for all $f \in \mathcal{E}^{\eta+\lambda}$, where $\lambda > 0$ is arbitrary.

Proof. In the Fourier expansion of f, we have $c_{\underline{0}} = \int_{E^k} f(\underline{t})d\underline{t}$. In the sequel, a summation symbol of the form $\sum^*$ means deletion of the origin from the range of summation. We have

$$\left|\frac{1}{N}\sum_{n=1}^{N} f(n\underline{\alpha}) - \int_{E^k} f(\underline{t})d\underline{t}\right|$$

$$= \left|\frac{1}{N}\sum_{\underline{h}}{}^{*} c_{\underline{h}} \sum_{n=1}^{N} e^{2\pi i<\underline{h},n\underline{\alpha}>}\right|$$

$$\leq \frac{1}{N}\sum_{\underline{h}}{}^{*} |c_{\underline{h}}| \left|\sum_{n=1}^{N} e^{2\pi i n<\underline{h},\underline{\alpha}>}\right| .$$

Note that $<\underline{h},\underline{\alpha}>$ is not an integer whenever $\underline{h} \neq \underline{0}$. As in Lemma 3.1, we obtain

$$\left|\sum_{n=1}^{N} e^{2\pi i n<\underline{h},\underline{\alpha}>}\right| \leq \frac{1}{2\|<\underline{h},\underline{\alpha}>\|} .$$

Using also the condition on the Fourier coefficients of f, we get

$$\left|\frac{1}{N}\sum_{n=1}^{N} f(n\underline{\alpha}) - \int_{E^k} f(\underline{t})d\underline{t}\right|$$

$$\leq \frac{1}{2N}\sum_{\underline{h}}{}^{*} |c_{\underline{h}}| \; \|<\underline{h},\underline{\alpha}>\|^{-1}$$

$$\leq \frac{M}{2N}\sum_{\underline{h}}{}^{*} R^{-\eta-\lambda}(\underline{h}) \; \|<\underline{h},\underline{\alpha}>\|^{-1} .$$

It suffices then to show that the sum on the right-hand side converges. By the inequality (2) in [44], we have

$$\sum_{\underline{h}}^{*} R^{-\eta-\lambda}(\underline{h})\,\|\langle\underline{h},\underline{\alpha}\rangle\|^{-1}$$

$$< (\eta+\lambda)^{k} \sum_{n_1,\ldots,n_k=1}^{\infty} (n_1\cdots n_k)^{-\eta-\lambda-1} \sum_{\substack{\underline{h}=(h_1,\ldots,h_k)\\ |h_j|\leq n_j}}^{*} \|\langle\underline{h},\underline{\alpha}\rangle\|^{-1}.$$

By the assumption on the type of $\underline{\alpha}$, there exists $C > 0$ such that

$$R^{\eta+\frac{\lambda}{3}}(\underline{h})\,\|\langle\underline{h},\underline{\alpha}\rangle\| \geq C \quad \text{for all} \quad \underline{h} \neq \underline{0}\,.$$

One can then show as in [44] (see also the proof of Theorem 6.1) that

$$\sum_{\substack{\underline{h}=(h_1,\ldots,h_k)\\ |h_j|\leq n_j}}^{*} \|\langle\underline{h},\underline{\alpha}\rangle\|^{-1} \leq L(n_1\cdots n_k)^{\eta+\frac{2\lambda}{3}}$$

with a constant L independent of $n_1,\ldots,n_k$. We finally obtain

$$\sum_{\underline{h}}^{*} R^{-\eta-\lambda}(\underline{h})\,\|\langle\underline{h},\underline{\alpha}\rangle\|^{-1}$$

$$\leq L(\eta+\lambda)^{k} \sum_{n_1,\ldots,n_k=1}^{\infty} (n_1\cdots n_k)^{-1-\frac{\lambda}{3}}$$

$$= L(\eta+\lambda)^k(\zeta(1+\tfrac{\lambda}{3}))^k ,$$

where ζ denotes the Riemann zeta-function. This completes the proof.

We infer from the above theorem that, depending on the degree of regularity of the integrand, we have a wider or smaller choice of $\underline{\alpha}$'s to obtain the error term $O(N^{-1})$. For instance, if f possesses partial derivatives of all orders, then any $\underline{\alpha}$ of finite type will do.

As to suggested choices of $\underline{\alpha}$, the same remarks as in section 6 apply. Favorable k-tuples seem to be those of the form $\underline{\alpha} = (\alpha_1,\ldots,\alpha_k)$, where the α_i are real algebraic numbers with 1, $\alpha_1,\ldots,\alpha_k$ linearly independent over the rationals (used in [44]), and $\underline{\alpha} = (e^{r_1},\ldots e^{r_k})$ with distinct nonzero rationals r_i (used in [70].)

Theorem 8.1, when compared with the error estimates for the method of low discrepancy sequences, seems to indicate that the Fourier series method is superior to the low discrepancy method. However, this statement can only be accepted with prudence. Firstly, one should bear in mind the fact that the error estimates in the low discrepancy method hold under extremely modest regularity assumptions (e.g., Riemann integrability, continuity, or bounded variation in the sense of Hardy and Krause). Secondly, the low discrepancy method works as well for integration domains other than E^k (see section 7), whereas there is no obvious way to extend the Fourier series method to other integration domains (if one were to try to replace the integrand f by fc_M, where c_M is the characteristic function of the integration domain M, this would immediately destroy the regularity conditions). Lastly, the Fourier series method has this built-in obstacle of the periodicity requirement on which, unfortunately, everything hinges. To get around this requirement involves an additional amount of computing time which is not negligible. In summary, one can say that the Fourier series method is favorable for integrals over E^k with an integrand having a high degree of regularity. For integrals over E^k with integrands having a low degree of regularity, and for integrals over other integration domains which cannot be transformed by a simple substitution into an integral over E^k, the method of low discrepancy sequences is recommended.

Now to the important problem of circumventing the periodicity condition on the integrand in the Fourier series method. The strategy will be to transform the given integrand f on E^k, by using changes of variables or algebraic manipulations, into a new integrand which satisfies the periodicity requirement on E^k, and which can then be extended periodically to $\mathbb{R}^k$. Ideally, these manipulations should preserve regularity properties of f. An easy way to reach these goals is to define

$$g(t_1,\ldots,t_k) = 2^{-k} \sum_{\substack{\varepsilon_1,\ldots,\varepsilon_k \\ \varepsilon_j=0,1}} f(\varepsilon_1+(-1)^{\varepsilon_1} t_1,$$

$$\ldots,\varepsilon_k + (-1)^{\varepsilon_k} t_k) \quad \text{for} \quad (t_1,\ldots,t_k)\in E^k.$$

Then g is obviously identical on opposite faces of E^k. Furthermore, we have

$$\int_{E^k} f(\underline{t})d\underline{t} = \int_{E^k} g(\underline{t})d\underline{t} \ ,$$

and the transition from f to g preserves regularity properties such as the existence and continuity of certain partial derivatives. More sophisticated methods have also been devised (see [23], [27], [35]). Zaremba [67] gives a detailed discussion of these methods and proposes one of his own which he supports by numerical evidence.

9. Improvement of the Method.

Returning again to the statement of Theorem 8.1, the following fact should be observed. Suppose we choose an $\underline{\alpha}$ of type η, and let $f\in\mathcal{E}^{\gamma}$ with γ much larger than η. Then the method, as it stands, does not honor the high degree of regularity of f. In other words, the theoretical error estimate for this f is the same as for an integrand which would just barely qualify for the method, i.e., for an integrand in

ε^{δ} with δ only slightly larger than η. This is certainly unjust.

How can we remedy this situation? The root of the problem is that our equal-weight formula for the integral $\int_{E_k} f(\underline{t})d\underline{t}$ seems to be too insensitive to detect any additional regularity of f. We shall therefore devise other types of weight distributions which better fit our purposes. A step in this direction was already undertaken by Haselgrove [23] (see also Haber [15]). Optimally, his choices of weights produce error terms of the form $O(N^{-4})$ for sufficiently regular integrands, when using essentially $4N$ nodes. By a simpler and more systematic method, we can prove the following result. For any positive integer s, one can obtain an error term $O(N^{-s})$ for sufficiently regular integrands if one uses essentially sN nodes and the weight distribution to be described in the sequel.

For given positive integers s and N, we define now weights of the form $a_{Nn}^{(s)}N^{-s}$ with $0 \leq n \leq s(N-1)$. The $a_{Nn}^{(s)}$ are positive integers which are introduced most conveniently by means of a generating function. Namely, the $a_{Nn}^{(s)}$ are determined from the polynomial identity

$$(9.1) \qquad \left(\sum_{j=0}^{N-1} z^j\right)^s = \sum_{n=0}^{s(N-1)} a_{Nn}^{(s)} z^n$$

over the field of complex numbers. Thus, for instance,

$$a_{Nn}^{(1)} = 1 \quad \text{for} \quad 0 \leqslant n \leqslant N-1,\ N \geqslant 1;$$

$$a_{Nn}^{(2)} = N-|N-1-n| \quad \text{for} \quad 0 \leqslant n \leqslant 2(N-1),\ N \geqslant 1.$$

With these weights, the following result holds.

<u>THEOREM 9.1</u>. Let s be a positive integer, and let $\underline{\alpha}$ be a k-tuple of irrationals of type η. Then we have

$$N^{-s} \sum_{n=0}^{s(N-1)} a_{Nn}^{(s)} f(n\underline{\alpha}) - \int_{E^k} f(\underline{t})d\underline{t} = O(N^{-s})$$

for all $f \in \mathcal{E}^{s\eta+\lambda}$, where $\lambda > 0$ is arbitrary.

<u>Proof.</u> We have

$$f(\underline{t}) = \sum_{\underline{h}} c_{\underline{h}}\, e^{2\pi i \langle \underline{h},\underline{t}\rangle} \quad \text{for all} \quad t \in R^k ,$$

and also

$$c_{\underline{0}} = \int_{E^k} f(\underline{t})d\underline{t} .$$

Therefore

$$N^{-s} \sum_{n=0}^{s(N-1)} a_{Nn}^{(s)} f(n\underline{\alpha}) - \int_{E^k} f(\underline{t})d\underline{t}$$

$$= N^{-s} \sum_{n=0}^{s(N-1)} a_{Nn}^{(s)} \sum_{\underline{h}} c_{\underline{h}}\, e^{2\pi i n\langle \underline{h},\underline{\alpha}\rangle} - c_{\underline{0}}$$

$$= N^{-s} \sum_{\underline{h}} c_{\underline{h}} \sum_{n=0}^{s(N-1)} a_{Nn}^{(s)}\, e^{2\pi i n\langle \underline{h},\underline{\alpha}\rangle} - c_{\underline{0}} .$$

Setting $z = 1$ in (9.1), we obtain

$$\sum_{n=0}^{s(N-1)} a_{Nn}^{(s)} = N^s .$$

Consequently, we arrive at

$$\begin{aligned} &N^{-s}\sum_{n=0}^{s(N-1)} a_{Nn}^{(s)} f(n\underline{\alpha}) - \int_{E^k} f(\underline{t})d\underline{t} \\ (9.2)\qquad &= N^{-s} \sum_{\underline{h}}{}^{*} c_{\underline{h}} \sum_{n=0}^{s(N-1)} a_{Nn}^{(s)} e^{2\pi i n<\underline{h},\underline{\alpha}>} . \end{aligned}$$

Each inner sum on the right hand side of (9.2) can be transformed by using (9.1) with $z = e^{2\pi i<\underline{h},\underline{\alpha}>}$. This yields

$$N^{-s}\sum_{n=0}^{s(N-1)} a_{Nn}^{(s)} f(n\underline{\alpha}) - \int_{E^k} f(\underline{t})d\underline{t}$$

$$= N^{-s} \sum_{\underline{h}}{}^{*} c_{\underline{h}} \Big(\sum_{j=0}^{N-1} e^{2\pi i j<\underline{h},\alpha>}\Big)^s .$$

Using $f \in \mathcal{E}^{s\eta+\lambda}$ and estimates as in the proof of Theorem 8.1, we infer that

$$\Big|N^{-s}\sum_{n=0}^{s(N-1)} a_{Nn}^{(s)} f(n\underline{\alpha}) - \int_{E^k} f(\underline{t})d\underline{t}\Big|$$

$$\leq M\, 2^{-s} N^{-s} \sum_{\underline{h}}{}^{*} R^{-s\eta-\lambda}(\underline{h}) \|<\underline{h},\underline{\alpha}>\|^{-s} .$$

To conclude the proof, it therefore suffices to show that the sum on the right hand side converges. Since $\underline{\alpha}$ is of type η, we have for some $C > 0$ that

$$R^{(\eta+\frac{\lambda}{s})(s-1)}(\underline{h})\,\|\langle\underline{h},\underline{\alpha}\rangle\|^{s-1} \geq C \quad \text{for all} \quad \underline{h} \neq \underline{0} .$$

Therefore

$$\sum_{\underline{h}}{}^{*} R^{-s\eta-\lambda}(\underline{h})\,\|\langle\underline{h},\underline{\alpha}\rangle\|^{-s}$$

$$\leq C^{-1} \sum_{\underline{h}}{}^{*} R^{-\eta-\frac{\lambda}{s}}(\underline{h})\,\|\langle\underline{h},\underline{\alpha}\rangle\|^{-1}.$$

But sums of the form appearing on the right hand side were already shown to be convergent in the proof of Theorem 8.1, and so we are done.

<u>COROLLARY 9.1</u>. If $\underline{\alpha} = (\alpha_1,\ldots,\alpha_k)$, where the α_i are real algebraic numbers such that 1, $\alpha_1,\ldots,\alpha_k$ are linearly independent over the rationals, and if $f \in \mathcal{E}^{\gamma}$ for some $\gamma > s$, then

$$N^{-s}\sum_{n=0}^{s(N-1)} a_{Nn}^{(s)} f(n\underline{\alpha}) - \int_{E^k} f(\underline{t})d\underline{t} = O(N^{-s}) .$$

These results are improvements on Haselgrove's results in [23] in two respects. First of all, they hold for all s, not only for $1 \leq s \leq 4$. Also, they exhibit simple and natural classes of $\underline{\alpha}$'s which work, rather than being mere existence theorems as in [23]. Corollary 9.1 holds of course as well with Baker's k-tuples (compare with section 6).

Finally, let us point out that the averages used in this improved method can be computed in a very economical way. This is due to the following simple identity.

LEMMA 9.1. For any s and N and any complex numbers $w_0,\dots,w_{s(N-1)}$, we have

$$\sum_{n=0}^{s(N-1)} a_{Nn}^{(s)} w_n = \sum_{j_1,\dots,j_s=0}^{N-1} w_{j_1+\dots+j_s} . \tag{9.3}$$

Proof. From (9.1) we see that $a_{Nn}^{(s)}$ is nothing else but the number of ways of representing n as a sum of integers $j_1,\dots,j_s$ with $0 \leq j_r \leq N-1$ for $1 \leq r \leq s$, regarding their order. Hence $a_{Nn}^{(s)}$ is identical with the number of occurrences of w_n on the right hand side of (9.3).

It is clear that, no matter whether we increase s or N, we can always use all the function values previously calculated. In essence, one just has to keep track of sums of the type $\sum_{j=0}^{N-1} w_{r+j}$. This feature of the present method is its major advantage over the method of good lattice points (see [69]), where the nodes depend on the a priori chosen value of N.

10. Effective Estimates.

When analyzing the proofs of Theorem 6.1, 8.1, and 9.1, one realizes that the only intractable constant is the constant $C = C(\underline{\alpha},\varepsilon)$ appearing in diophantine inequalities such as

$$R^{\eta+\varepsilon}(\underline{h})\,\|\langle\underline{h},\underline{\alpha}\rangle\| \geq C \quad \text{for all } \underline{h} \neq \underline{0}. \tag{10.1}$$

Therefore, explicit error bounds for the Fourier series method (and for the method of low discrepancy sequences in the case $k \geq 2$) have to rely solely on the so-called effective methods in diophantine approximations which have as their object the derivation of inequalities such as (10.1) with a computable constant C. These effective methods have received much attention during the last decade. The pioneering work of Baker is widely known by now, its appeal being enormously enhanced by the dramatic impact it had on the class-number problem.

For the important case where $\underline{\alpha} = (\alpha_1,\ldots,\alpha_k)$ is a k-tuple of real algebraic numbers with $1,\alpha_1,\ldots,\alpha_k$ linearly independent over the rationals, it seems to be extremely difficult to get nontrivial effective estimates as soon as $k \geq 2$. There is an easy and obvious method using the norm function relative to the field extension $Q(\alpha_1,\ldots,\alpha_k)$ of the field Q of rationals. This was for instance carried out in [44] with the special values $\alpha_j = 2^{j/(k+1)}$ for $1 \leq j \leq k$, yielding the inequality

$$\left(\frac{9k}{2}R(\underline{h})\right)^k \,\|\langle\underline{h},\underline{\alpha}\rangle\| \geq 1 \quad \text{for all} \quad \underline{h} \neq \underline{0} .$$

The large exponent k of $R(\underline{h})$ is of course very unsatisfactory. However, this value of the exponent is a sort of natural barrier, in the sense that any improvement of this exponent will require much deeper methods.

In the multidimensional case, such methods have only been developed for special circumstances. We refer to Baker [4] and Osgood [47]. C.J. Moreno has pointed out to me that the result of Baker [3] on the k-tuples $(e^{r_1},\ldots,e^{r_k})$ with distinct nonzero rationals r_i could be made effective by using methods of his paper [39]. In the one-dimensional problem concerning algebraic irrationals, the results of Baker [1], [2] are available. Other results in this direction are contained in the papers of Davenport [10] and Osgood [48].

REFERENCES

[1] A.Baker, Rational approximations to $\sqrt[3]{2}$ and other algebraic numbers, Quart. J. Math. (Oxford) (2) 15, 375-383, (1964).

[2] A.Baker, Rational approximations to certain algebraic numbers, Proc. London Math. Soc. (3) 14, 385-398, (1964).

[3] A.Baker, On some diophantine inequalities involving the exponential function, Canad. J. Math. 17, 616-626, (1965).

[4] A.Baker, Simultaneous rational approximations to certain algebraic numbers, Proc. Cambridge Phil. Soc. 63, 693-702, (1967).

[5] H. Behnke, Über die Verteilung von Irrationalitäten mod 1, Abh. Math. Sem. Hamburg 1, 252-267, (1922).

[6] H. Behnke, Zur Theorie der diophantischen Approximationen, I, Abh. Math. Sem. Hamburg 3, 261-318, (1924).

[7] P. Billingsley and F. Topsøe, Uniformity in weak convergence, Z. Wahrscheinlichkeitstheorie 7, 1-16, (1967).

[8] T. Bonnesen and W. Fenchel, Theorie der konvexen Körper, Ergebn. Math. Grenzgeb., Band 3, Heft 1, Springer,Berlin, (1934).

[9] H. Davenport, Note on irregularities of distribution, Mathematika 3, 131-135,(1956).

[10] H. Davenport, A note on Thue's Theorem, Mathematika 15, 76-87, (1968).

[11] P.J. Davis and P. Rabinowitz,Numerical integration, Blaisdell Publishing Co., Waltham, Mass., 1967.

[12] P. Erdős and P. Turán, On a problem in the theory of uniform distribution I, Indag. Math. 10, 370-378,(1948).

[13] M. Fréchet, Généralités sur les probabilités, Éléments aléatoires, Deuxième éd., Gauthier-Villars, Paris, 1950.

[14] S. Haber, On a sequence of points of interest for numerical quadrature, J. Res. Nat. Bur. Standards Sect. B70, 127-136, (1966).

[15] S. Haber, Numerical evaluation of multiple integrals, SIAM Rev. 12, 481-526,(1970).

[16] J.H. Halton, On the efficiency of certain quasi-random sequences of points in evaluating multidimensional integrals, Numer. Math. 2, 84-90, (1960).

[17] J.H. Halton, A retrospective and prospective survey of the Monte Carlo method, SIAM Rev. 12, 1-63, (1970).

[18] J.H. Halton, Estimating the accuracy of quasi-Monte Carlo integration, Proc. Sympos. on Applications of Number Theory to Numerical Analysis (Montreal, 1971). Edited by S.K. Zaremba, Academic Press, New York, 1972.

[19] J.H. Halton and S.K. Zaremba, The extreme and L^2 discrepancies of some plane sets, Monatsh. Math. 73, 316-328, (1969).

[20] J.M. Hammersley and D.C. Handscomb, Monte Carlo methods, Methuen, London, 1964.

[21] G.H. Hardy and J.E. Littlewood, Some problems of Diophantine approximation: The lattice points of a right-angled triangle, I., Proc. London Math. Soc. 20, 15-36, (1922).

[22] G.H. Hardy and J.E. Littlewood,Some problems of Diophantine approximation: The lattice points of a right-angled triangle, II.,Abh. Math. Sem. Hamburg 1, 212-249, (1922).

[23] C.B. Haselgrove, A method for numerical integration, Math. Comp. 15, 323-337, (1961).

[24] E. Hecke, Über analytische Funktionen und die Verteilung von Zahlen mod. Eins, Abh. Math. Sem. Hamburg 1, 54-76, (1922).

[25] G. Helmberg, Gleichverteilte Folgen in lokal kompakten Räumen, Math. Z. 86, 157-189, (1964).

[26] E. Hlawka, Funktionen von beschränkter Variation in der Theorie der Gleichverteilung, Ann. Mat. Pura Appl. (IV) 54, 325-333, (1961).

[27] E. Hlawka, Zur angenäherten Berechnung mehrfacher Integrale, Monatsh. Math. 66, 140-151, (1962).

[28] E. Hlawka, Discrepancy and uniform distribution of sequences, Compositio Math. 16, 83-91, (1964).

[29] E. Hlawka, Uniform distribution modulo 1 and numerical analysis, Compositio Math. 16, 92-105, (1964).

[30] E. Hlawka, Discrepancy and Riemann integration, Studies in Pure Mathematics, pp. 121-129. Edited by L. Mirsky. Academic Press, New York, 1971.

[31] E. Hlawka and R. Mück, A transformation of equidistributed sequences, Proc. Sympos. on Applications of Number Theory to Numerical Analysis (Montreal, 1971). Edited by S.K. Zaremba, Academic Press, New York, 1972.

[32] J.F. Koksma, Een algemeene stelling uit de theorie der gelijkmatige verdeeling modulo 1, Mathematica (Zutphen B) 11, 7-11, (1942-43).

[33] J.F. Koksma, Eenige integralen in de theorie der gelijkmatige verdeeling modulo 1, Mathematica (Zutphen B)11, 49-52, (1942-43).

[34] J.F. Koksma, Some theorems on diophantine inequalities, Math. Centrum Amsterdam, Scriptum No. 5, 1950.

[35] N.M. Korobov, Number-theoretical methods in approximate analysis (Russian), Fizmatgiz, Moscow, 1963.

[36] L. Kuipers, Simple proof of a theorem of J.F. Koksma, Nieuw Tijdschrift voor Wisk. 55, 108-111, (1967).

[37] L. Kuipers and H. Niederreiter, Uniform distribution of sequences,(Wiley and Sons).

[38] W.J. Leveque, An inequality connected with Weyl's criterion for uniform distribution, Proc. Symp. Pure Math. VIII, pp. 22-30. American Mathematical Society, Providence, R. I., 1965.

[39] C.J. Moreno, Values of exponential polynomials at algebraic points, to appear.

[40] H. Niederreiter, Diskrepanz in kompakten abelschen Gruppen II, Manuscripta Math. 1, 293-306, (1969).

[41] H. Niederreiter, Almost-arithmetic progressions and uniform distribution, Trans. Amer. Math. Soc. 161, 283-292, (1971).

[42] H. Niederreiter, Discrepancy and convex programming, Ann. Mat. Pura Appl., to appear.

[43] H. Niederreiter, Methods for estimating discrepancy, Proc. Sympos. on Applications of Number Theory to Numerical Analysis (Montreal, 1971). Edited by S.K. Zaremba. Academic Press, New York, 1972.

[44] H. Niederreiter, On a number-theoretical integration method, Aequationes Math., to appear.

[45] H. Niederreiter, Metric theorems on the distribution of sequences, Proc. Symp. Pure Math. XXIV. American Mathematical Society, Providence, R. I., to appear.

[46] H. Niederreiter, Irregularities of distribution, in preparation.

[47] C.F. Osgood, The simultaneous diophantine approximation of certain k-th roots, Proc. Cambridge Phil. Soc. 67, 75-86, (1970).

[48] C.F. Osgood, The diophantine approximation of roots of positive integers, J. Res. Nat. Bur. Standards Sect. B74, 241-244, (1970).

[49] A. Ostrowski, Bemerkungen zur Theorie der Diophantischen Approximationen I, Abh. Math. Sem. Hamburg 1, 77-98, (1922).

[50] L.G. Peck, On uniform distribution of al-

gebraic numbers, Proc. Amer. Math. Soc. 4, 440-443, (1953).

[51] W. Philipp, Empirical distribution functions and uniform distribution mod 1, this volume.

[52] R.D. Richtmyer, On the evaluation of definite integrals and a quasi-Monte Carlo method based on properties of algebraic numbers, Report LA-1342, Los Alamos Sci. Lab., Los Alamos, N.M., 1952.

[53] K.F. Roth, On irregularities of distribution, Mathematika 1, 73-79, (1954).

[54] W.M. Schmidt, Simultaneous approximation to algebraic numbers by rationals, Acta Math. 125, 189-201, (1970).

[55] W.M. Schmidt, Irregularities of distribution. VII, Acta Arith., to appear.

[56] I.M. Sobol', An exact estimate of the error in multidimensional quadrature formulae for functions of the classes $\tilde{W}_1$ and $\tilde{H}_1$ (Russian), Ž. Vyčisl. Mat. i Mat. Fiz. 1, 208-216, (1961); Eng. trans. in U.S.S.R. Comp. Math. and Math. Phys. 1, 228-240, (1962).

[57] I.M. Sobol', An accurate error estimate for multidimensional quadrature formulae for the functions of the class S_p (Russian), Dokl. Akad. Nauk SSSR 132, 1041-1044 (1960); Eng. trans. in Soviet Math. Doklady 1, 726-729, (1961).

[58] I.M. Sobol', On the evaluation of multidimensional integrals (Russian), Dokl. Akad. Nauk SSSR 139, 821-823, (1961); Eng. trans. in Soviet Math. Doklady 2, 1022-1025, (1961).

[59] I.M. Sobol', The distribution of points in a cube and the approximate evaluation of integrals (Russian), Ž. Vyčisl. Mat. i Mat. Fiz. 7, 784-802, (1967); Eng. trans. in U.S.S.R. Comp. Math. and Math Phys. 7, No. 4, 86-112, (1967).

[60] J.G. Van der Corput and C. Pisot, Sur la discrèpance modulo un, Indag. Math. 1, 143-153, 184-195, 260-269, (1939).

[61] T.T. Warnock, Computational investigations of low-discrepancy point sets, Proc. Symp.

on Applications of Number Theory to Numerical Analysis (Montreal, 1971), pp. 319-344. Edited by S.K. Zaremba. Academic Press, New York, 1972.

[62] H. Weyl, Über die Gleichverteilung von Zahlen mod. Eins., Math. Ann. 77, 313-352, (1916).

[63] A. Yudin, New proof of Erdös-Turan's theorem (Russian), Litovsk. Mat. Sb. 9, 839-848, (1969).

[64] V.V. Yurinskii, On inequalities for large deviations for certain statistics, Theor. Probability Appl. 16, 385-387, (1971).

[65] S.K. Zaremba, Good lattice points, discrepancy, and numerical integration, Ann. Mat. Pura Appl. (IV) 73, 293-317, (1966).

[66] S.K. Zaremba, Some applications of multidimensional integration by parts, Ann. Polon. Math. 21, 85-96, (1968).

[67] S.K. Zaremba, A quasi-Monte Carlo method for computing double and other multiple integrals, Aequationes Math. 4, 11-22, (1970).

[68] S.K. Zaremba, La discrépance isotrope et l'intégration numérique, Ann. Mat. Pura Appl. (IV) 87, 125-136, (1970).

[69] S.K. Zaremba, La méthode des "bons treillis" pour le calcul numérique des intégrales multiples, Proc. Symp. on Applications of Number Theory to Numerical Analysis (Montreal 1971), pp. 39-119. Edited by S.K. Zaremba, Academic Press, New York, 1972.

[70] P. Zinterhof, Einige zahlentheoretische Methoden zur numerischen Quadratur und Interpolation, Sitzungsber. Österr. Akad. Wiss., Math. - Naturw. Kl., Abt. II, 177, 51-77, (1969).

Southern Illinois University, Carbondale, Ill.

A DIOPHANTINE PROPERTY OF $J_o(z)$

Charles F. Osgood

Let Ω denote the field of algebraic numbers. Let y denote <u>any</u> nonzero function which satisfies a linear homogeneous differential equation with coefficients in $\Omega[z]$ and which may be written, near $z = 0$, in the form $y = \sum_{j=1}^{m} z^{c_j} f_j(z)$, where each $c_j \geq 0$ is a real number and each $f_j(z)$ is analytic at $z = 0$. Let

$$(y * J_o)(z) \overset{\text{def.}}{\equiv} y(z) * J_o(z)$$

denote $\int_o^z y(t)J_o(z-t)dt$, which is defined on (at least) D, the largest open disk about $z = 0$ on which each f_j is analytic.

<u>THEOREM</u>. At all points z_o of $\Omega \cap D$, with but at most a finite number of exceptions, there exist positive integers N, λ, and K, depending on both $y(z)$ and z_o such that

$$\max_{0 \leq j \leq N} \{|y^{(j)}(z_o) - p_j q^{-1}| + |(y * J_o)^{(j)}(z_o) - r_j q^{-1}|\} \geq \lambda |q|^{-K}$$

for all Gaussian integers $p_o, \ldots, p_N$, $r_o, \ldots, r_N$ and all nonzero Gaussian integers q.

We recall that

$$(z^{\alpha}(\Gamma(\alpha+1))^{-1}) * (z^{\beta}(\Gamma(\beta+1))^{-1})$$
$$= z^{\alpha+\beta+1}(\Gamma(\alpha+\beta+2))^{-1} ,$$

so it is possible to write out $(y * J_o)(z)$ in a series similar to that for y(z). One could also carry out the following proof using $J_o(\gamma z)$, instead of $J_o(z)$, for any nonzero $\gamma \in Q(i)$. In particular, choosing $\gamma = 2i$, we see that if

$$y(z) = \sum_{n=0}^{\infty} a_n (n!)^{-1} z^n$$

then convoluting by $J_o(2iz)$ we obtain $\sum_{n=0}^{\infty} b_n (n!)^{-1} z^n$ where

$$b_n = \sum_{m \geq 0} a_{n-2m-1} \binom{2m}{m}.$$

<u>Proof</u>. We start with the two familiar recurrence relations for $J_n(z)$, see page 17 of [3],

(1) $$J_{n-1}(z) + J_{n+1}(z) = 2nz^{-1}J_n(z)$$

and

(2) $$J_{n-1}(z) - J_{n+1}(z) = 2J_n'(z) .$$

We next define two linear operators S and M on the vector space generated over C by all formal series of the form $\sum_{n=0}^{\infty} a_{n+c} J_{n+c}(z) t^{n+c}$ for all non-negative real numbers c and all complex numbers a_{n+c}. Set $M = t \frac{\partial}{\partial t}$ and let

$$S(J_{n+c}(z)t^{n+c}) = J_{n-1+c}(z)t^{n-1+c} .$$

From (1) and (2) above we see that

$$M = t\frac{\partial}{\partial t} = \frac{1}{2}z(t^{-1}S^{-1} + tS)$$

and

$$z\frac{\partial}{\partial z} = \frac{1}{2}z(-t^{-1}S^{-1} + tS).$$

Then

(3) $$S = (tz)^{-1}(z\frac{\partial}{\partial z} + t\frac{\partial}{\partial t}) ,$$

(4) $$M = \frac{1}{2}z(t^{-1}S^{-1} + tS) ,$$

and

(5) $$S^{-1} = tz^{-1}(t\frac{\partial}{\partial t} - z\frac{\partial}{\partial z}) .$$

Let us suppose that for some $F(z,t)$ in our above vector space all of the coefficients a_{n+c} satisfy the recurrence relation

(6) $$\sum_{j\geqslant 0} p_j(n+c)a_{n+c+j} = 0 ,$$

where each $p_j(n)\in\Omega[n]$ and $p_0(n) \neq 0$. Then we see that we have the following operator equation for $F(z,t)$,

(7) $$\sum_{j\geqslant 0} p_j(M)S^j F(z,t) = 0 .$$

Also, for each positive integer N we have, using (4),

(8) $$\sum_{j\geqslant 0} p_j(\frac{1}{2}z(t^{-1}S^{-1}+tS)-N)S^{-N+j}F(z,t) = 0$$

Using our equation (5) we see that for all nonnegative integers a and b

$$(9) \qquad S^{-1}(t^a z^b) = t^a z^b(S^{-1} + (a-b)tz^{-1})$$

and

$$(10) \qquad t^a z^b S^{-1} = (S^{-1} + (b-a)tz^{-1})t^a z^b .$$

Using (9) and (10) to simplify (8) we see that, where $S^{-(N+\phi)}$ is the highest power of S^{-1} to appear in (8) the coefficient of $S^{-(N+\phi)}$ is

$$q(zt^{-1}) \overset{\text{def}}{=} \sum_{j\geq 0} b_j(\tfrac{1}{2}\, zt^{-1})^{\phi+j},$$

where b_j is the coefficient of $n^{\phi+j}$ in $p_j(n)$. (Since some $b_j \neq 0$ by assumption, we see that $q(zt^{-1}) \not\equiv 0$.) Now the

$$(q(zt^{-1}))^{N+\phi} S^{-(N+\phi)} F(z,t)$$

for $\phi = 0,1,\ldots$ generate a finitely generated module over $K[z,t,t^{-1},N]$ for some $K \subset \Omega$ with $[K : Q(i)] < \infty$. Then there exists a nonzero $P(zt^{-1}) \in Q[i,zt^{-1}]$ such that the $(P(zt^{-1}))^{N+\phi}$ $S^{-(N+\phi)}F(z,t)$ generate a finitely generated module over $Q[i,z,t,t^{-1},N]$. By the ascending chain condition on ideals in $Q[i,z,t,t^{-1},N]$ (henceforth written as a.c.c.) there exists a recurrence relation giving, for some $\phi_1 \geq 1$,

$$(p(zt^{-1}))^{N+\phi_1} S^{-(N+\phi_1)} F(z,t)$$

as a linear combination over

$$Q[i, z, t, t^{-1}, N]$$

of the

$$(p(zt^{-1}))^{N+\phi} S^{-(N+\phi)} F(z,t)$$

for all $0 \leq \phi < \phi_1$.

Consider $F(z,1)$. It is, formally, a Neuman series and, as such, converges at each point of a punctured disk centered at $z = 0$ to a differentiable function (see [3]). It is not difficult to see that the radius of convergence of the Neumann series is the same as the radius of convergence of

$$\sum_{n=0}^{\infty} a_{n+c} \left(\tfrac{1}{2}z\right)^{n+c} (\Gamma(n+1+c))^{-1} ,$$

(see [3] Chapter XVI).

Suppose that $F(z,1)$ has a nonzero radius of convergence about $z = 0$, that $z_o \in \Omega$ is not a zero of $zp(z)$, and that z_o is in the disk of convergence of $F(z,1)$. Then, for each non-negative integer n, $S^{-n}F(z_o,1)$, $S^nF(z_o,1)$, and $(t \frac{\partial}{\partial t})^n F(z_o,t)$, at $t = 1$, are defined. Also there exists a nonzero Gaussian integer K_1 such that the $K_1^{(N+\phi)} S^{-(N+\phi)} F(z_o,1)$ generate a finitely generated module over $Z[i,N]$. By the a.c.c. of $Z[i,N]$ we have that for some $\phi_2 \geq 0$ we may write

$$K_1^{N+\phi_2} S^{-(N+\phi_2)} F(z_o,1)$$

as a linear combination over $Z[i,N]$ of the $K_1^{N+\phi} S^{-(N+\phi)} F(z_o,1)$ for $0 \leq \phi \leq \phi_2$.

We wish to apply Lemma III of [2]. Set $F(s)$ there equal to $K_1^s S^{-s}F(z_0,1)$. Clearly

$$|K_1^s S^{-s}F(z_0,1)| \leq K_2^s(s!)^{-1}$$

for some $K_2 > 0$ independent of s, if $s \geq 1$. Thus, unless $S^{-s}F(z_0,1)$ is zero for all but a finite number of positive values of s, we have that for positive integers u, λ', and K

$$\max_{0\leq k\leq u} \{|K_1^k S^{-k}F(z_0,1) - p_k q^{-1}|\} \geq \lambda'|q|^{-K}, \tag{11}$$

for all nonzero Gaussian integers q and all Gaussian integers p_k.

If $c \geq 0$ is the smallest subscript such that $a_c \neq 0$ in the expansion of $F(z,t)$ then one may show that

$$\lim_{N\to\infty}(S^{-N}F(z_0,1))(\Gamma(N+1+c)(\tfrac{1}{2}z_0)^{N+c})^{-1}$$
$$= a_c \neq 0 .$$

(Here one needs the easily proven result that $\prod_{n=1}^{\infty}(1-a_n) = 0$ if (i) each $a_n \geq 0$, (ii) $\lim_{n\to\infty} a_n = 0$, and (iii) $\sum_{n=1}^{\infty} a_n = +\infty$.) We see that then, since it is impossible for all but a finite number of the $S^{-N}F(z_0,1)$ to be zero, (11) holds.

From (2) we see that

$$tS^{-(n-1)} - t^{-1}S^{-(n+1)} = 2\,\frac{\partial}{\partial z}\,S^{-n} ,$$

for each integer n. Thus we may write the $S^{-k}F(z_o,1)$ in (11) in terms of a finite number of partial derivatives of $F(z_o,1)$ and $S^{-1}F(z_o,1)$ with respect to z. Now by (5) we see that we may write

$$S^{-1}F(z,1) \quad \text{as} \quad z^{-1}\frac{\partial}{\partial t}F(z,1) - \frac{\partial}{\partial z}F(z,1) .$$

Hence, by elementary results in diophantine approximation, we may replace the $S^{-k}F(z_o,1)$ in (11) by a finite number of partial derivatives, with respect to z, of $F(z,1)$ and $z^{-1}\frac{\partial}{\partial t}F(z,1)$ at $z = z_o$, if we replace also λ' by a (possibly) larger constant $\frac{1}{2}\lambda$.

Suppose that we are given $y(z)$ as in the hypotheses of the Theorem. Then, see [3] Chapter XVI, we may expand $zy(z)$ in a Neuman series which has a radius of convergence equal to the radius of convergence of the original expansion for $y(z)$. Thus set

$$(12) \qquad zy(z) = \sum_{j=1}^{m}\sum_{n=0}^{\infty} b_{n+c_j} J_{n+c_j}(z) .$$

Now set

$$(13) \qquad F(z,t) = \sum_{j=1}^{m}\sum_{n=0}^{\infty} b_{n+c_j}(n+c_j)^{-1} J_{n+c_j}(z) t^{n+c_j} .$$

Then we see that we have $z^{-1} \frac{\partial}{\partial t} F(z,1) = y(z)$. In [1] on page 182 of Vol. I we learn that if $R(n+c_j) > 0$ the Laplace transform of $z^{-1}J_{n+c}(az)$ is

$$(n+c_j)^{-1} a^{n+c_j}(z + (z^2+a^2)^{1/2})^{-(n+c_j)},$$

while if $R(n + c_j) > -1$ the Laplace transform of $J_{n+c_j}(az)$ is

$$(z^2+a^2)^{-1/2} a^{n+c_j}(z + (z^2+a^2)^{1/2})^{-(n+c_j)}.$$

(These formulas hold if in the transformed functions $R(z) > |Im(a)|$.) Consider $y(z)*J_o(z)$. Since

$$y(z) = \sum_{j=1}^{m} \sum_{n=0}^{\infty} b_{n+c_j} (z^{-1}J_{n+c_j}(z))$$

we see that

$$y(z) * J_o(z) = \sum_{j=1}^{m} \sum_{n=0}^{\infty} b_{n+c_j} \Phi_{n+c_j} ,$$

where Φ_{n+c_j} is the inverse Laplace transform of

$$(n+c_j)^{-1}(z^2+1)^{-1/2}(z + (z^2+1)^{1/2})^{-(n+c_j)}.$$

Then $\Phi_{n+c_j} = (n+c_j)^{-1}J_{n+c_j}(z)$. (The condition that each $n+c_j > 0$ above holds since $zy(z)$ vanishes at $z = 0$.) Thus $F(z,1) = y(z) * J_o(z)$. This would prove the Theorem if we knew that the b_{n+c_j}'s satisfy a recurrence relation of type (6).

Consider the formal "power series" about $z = \infty$

$$\hat{F}_t(z,1) = (z^2+1)^{-1/2} \sum_{j=1}^{m} \sum_{n=0}^{\infty} b_{n+c_j} (z + (z^2+1)^{1/2})^{-(n+c_j)} .$$

Under the mapping which sends each $z^{-\theta}$ into $z^{\theta-1}(\Gamma(\theta))^{-1}$ we see that $\hat{F}_t(z,1)$ goes into $F_t(z,1)$. Set $g(z) = z(y(z))$. Then $\hat{g}(z) = \hat{F}_t(z,1)$, so $\hat{F}_t(z,1)$ satisfies a linear homogeneous differential equation with coefficients in $\Omega[z]$. Then so does

$$\sum_{j=1}^{m} \sum_{n=0}^{\infty} b_{n+c_j} (z + (z^2+1)^{1/2})^{-(n+c_j)}$$

and finally, so does

$$\sum_{j=1}^{m} \sum_{n=0}^{\infty} b_{n+c_j} z^{n+c_j} .$$

It follows that the b_{n+c_j}'s satisfy a recurrence relation of type (6). This proves the Theorem.

REFERENCES

[1] A. Erdélyi, The Bateman Manuscript Project Tables of Integral Transforms, McGraw-Hill, New York, 1954.

[2] C.F. Osgood, A method in diophantine Approximation (IV), Acta Arithmetica XVI (1969) pp. 23-40.

[3] G.N. Watson, A treatise in the theory of Bessel Functions, Cambridge University Press, 1966.

Mathematics Research Center, Naval Research Laboratory, Washington, D.C.

EMPIRICAL DISTRIBUTION FUNCTIONS AND UNIFORM DISTRIBUTION MOD 1

Walter Philipp

1. Introduction

Let $\langle x_n, n = 1,2,\ldots\rangle$ be a sequence of real numbers. Denote by $\{\varepsilon\}$ the fractional part of ε. For any interval I contained in the unit interval $U = [0,1)$ and with length $\lambda(I)$ denote by $A(N,I)$ the number of x_n, $n \leq N$ with $\{x_n\} \in I$. The sequence $\langle x_n\rangle$ is called uniformly distributed mod 1 if

$$\lim_{N\to\infty} \frac{1}{N} A(N,I) = \lambda(I) \tag{1}$$

for all intervals I contained in the unit interval. The discrepancy D_N of the sequence is defined as

$$D_N = \sup_{I\subset U} \left|\frac{1}{N} A(N,I) - \lambda(I)\right| . \tag{2}$$

It is easy to see that the sequence $\langle x_n\rangle$ is uniformly distributed mod 1 iff

$$\lim_{N\to\infty} D_N = 0 . \tag{3}$$

Without loss of generality we can and do assume that all x_n are contained in the unit interval $[0,1)$. We then can view the sequence $\langle x_n\rangle$ as a point in the infinite-dimensional unit cube $\Omega_\infty = \overset{\infty}{\underset{k=1}{X}} [0,1)$. If we put on Ω_∞ the infinite product measure λ_∞ together with the usual Lebesgue measurability we get a probability space $(\Omega_\infty, F_\infty, \lambda_\infty)$. It is well known (in fact it follows from the strong law of large numbers for independent identically distributed random variables) that (1) holds for almost all sequences, i.e. (1) and hence (3) holds for all sequences except a set of λ_∞-measure 0. But in view of Kolmogorov's existence theorem the last statement is just the Glivenko-Cantelli theorem which is usually formulated as follows: Let $X_1, X_2, \ldots$ be a sequence of independent random variables each having uniform distribution on the unit interval, i.e., $P(X \leq x) = 0$ if $x < 0$, $P(X \leq x) = x$ for $0 \leq x \leq 1$ and $P(X \leq x) = 1$ for $x > 1$. Let $A(N,x)$ be the number of indices $n \leq N$ with $X_n \leq x$. In analogy to (2) put

$$\Delta_N = \sup_{0 \leq x \leq 1} \left| \frac{1}{N} A(N,x) - x \right| .$$

Then

$$P(\lim_{N \to \infty} \Delta_N = 0) = 1 .$$

The limit distribution of $N^{1/2} \Delta_N$ has been established by Kolmogorov [9]:

$$(4) \quad \lim_{N \to \infty} P(N^{1/2} \Delta_N \leq \alpha) = \sum_{k=-\infty}^{\infty} (-1)^k e^{-2k^2 \alpha^2}, \quad \alpha \geq 0$$

Chung [3] gave estimates on the large deviations which enabled him to prove a law of the iterated logarithm:

$$(5) \qquad P\left\{ \limsup_{N\to\infty} \frac{N\,\Delta_N}{\sqrt{2N \log\log N}} = \frac{1}{2}\right\} = 1 .$$

Later, but independently Cassels [2] proved a result somewhat stronger than (5). (See Theorem 1 below). Chung's result (5) as well as Cassel's results have been extended to dimensions $r > 1$. The existence of the limit distribution for $N^{1/2}D_N$ for $r > 1$ has been established by Kiefer and Wolfowitz [8]. Depending on the point of view one wishes to adopt one considers sequences of points in U^r, the r-dimensional unit cube and states the theorems in terms of almost all sequences, or else one talks about sequences $<X_n>$ of independent r-dimensional random vectors. We shall use both conventions. In order to formulate the law of the iterated logarithm for discrepancies in r-space we adopt the following notation. Let $\underline{S}$ be a class of measurable sets $S \in \underline{S}$ contained in U^r and let $A(N,S)$ be the number of $n \leq N$ with $X_n \in S$. In analogy to (2) we define

$$(6) \qquad D_N(\underline{S}) = \sup_{S\in\underline{S}} \left|\frac{1}{N} A(N,S) - V(S)\right|$$

where $V(S)$ is the volume of S. We are interested in the law of the iterated logarithm in the form

$$(7) \qquad P\left\{\limsup_{N\to\infty} \frac{ND_N(\underline{S})}{\sqrt{2N \log\log N}} = \frac{1}{2}\right\} = 1$$

for different classes $\underline{S}$.

Of course, Chung's result (5) can be paraphrased as follows: (7) holds for $r = 1$ and $\underline{S} = \langle [0,x),\ 0 \leq x \leq 1 \rangle$; whereas Cassels' theorem yields (7) for $r = 1$ and $\underline{S} = \langle [\alpha,\beta),\ 0 \leq \alpha < \beta \leq 1 \rangle$. Kiefer [7] proved (7) for $r \geq 1$ and the class $\underline{S}$ of rectangles with one vertex at the origin. Modifying Cassels' method Zaremba [19] generalized one of Cassels' theorems to higher dimensions which implies (7) for $r \geq 1$ and the class of rectangles with sides parallel to the coordinate planes. Later, but independently, the author [14] proved (7) for $r > 1$ and general rectangles.

In this paper we shall prove (7) for $r \geq 1$ and the class of ellipsoids with axes parallel to the coordinate planes and for $r = 2$ and the class of convex sets. In other words for $r = 2$ we prove the law of the iterated logarithm for isotrope discrepancies. Actually, we shall give two proofs for the latter one. One of them fails for dimension $r > 2$. The other one leads to a combinatorial problem which I was unable to solve. As it will be clear from the proofs, the theorem continues to hold for more general than convex sets, for example for sets whose boundaries have slopes which do not vary too quickly. The results should, perhaps, be compared with theorems of Schmidt [17] and Zaremba [20]. Schmidt [17] proved that $ND_N(\underline{S}) >> N^{1/2-1/(2r)-\varepsilon}$ for any r-dimensional sequence and the class $\underline{S}$ of spheres, whereas Zaremba [20] proved the same result without the ε in the exponent for the class $\underline{S}$ of convex sets. The failure of one of the methods for $r > 2$ is in accordance with a conjecture of Niederreiter [12]. This conjecture states that $ND_N(\underline{S}) > N^{1-\frac{1}{r}}$ infinitely often for any r-dimensional sequence and the class $\underline{S}$ of convex sets.

The case of dependent random variables has also been studied extensively. For Kolmogorov type results see Billingsley [1] where actually invariance principles for empirical distribution functions are proved. A law of the iterated logarithm for sequences of the form $\{2^{\ell_n}x\}$ was proved by S. and L. Gaal [5]. This result was subsequently generalized and improved upon in [13] and [14]. In [14] a rather general theorem was given which includes all the log log-laws mentioned above. Recently Stackelberg [18] used it to prove a log log-law for C-uniform distribution. In section 2 we shall slightly improve this general theorem in the case of independent random variables. Applications are given in the subsequent sections.

For a comprehensive account on these and similar results, see Niederreiter [11].

I am grateful to Professor L. Kiefer for calling his paper [7] to my attention and to Professor P. Turán for pointing out to me that Lemma 3 below is due to Fejes Toth [4].

2. A Uniform Log Log Law

THEOREM 1. Let $A = \langle z_n(\alpha),\ n = 1,2,\ldots,\ \alpha \in A\rangle$ be a family of sequences of random variables. Suppose that for each non-negative integer t there is a class A_t whose elements we denote by $\langle x_n(t),\ n = 1,2,\ldots\rangle$ with the following properties:

(i) $$1 \leq \text{card } A_t \leq \exp(\rho(t))$$

where $\rho(t) > t^{6/\eta}$ is monotone in t and $0 < \eta < 1$ is a constant.

(ii) Each sequence $\langle x_n(t),\ n = 1,2,\ldots\rangle$ is a sequence of independent identically distributed random variables uniformly bounded by 1 (say) with $E(x_n^2(t)) \ll (\rho(t))^{-1-\eta}$ uniformly in t.

(iii) For each random variable $z_n(\alpha)$ and for any $s \geq 0$ there are two sequences of random variables $x_n(0), x_n(1), \ldots,\ ;\ \bar{x}_n(0), \bar{x}_n(1), \ldots$ such that

$$\bar{y}_n(s) \overset{\text{def.}}{=} \sum_{t=0}^{s} \bar{x}_n(t) \leq z_n(\alpha) \leq \sum_{t=0}^{s} x_n(t) \overset{\text{def.}}{=} y_n(s)$$

and

$$0 \leqslant \sum_{t=0}^{s} E(x_n(t) - \bar{x}_n(t)) \leqslant (\rho(s))^{-1-\eta}$$

where $<x_n(t),\ n = 1,2,\ldots>$, $<\bar{x}_n(t),\ n = 1,2,\ldots>$ $\in A_t$ for $t = 0,1,2,\ldots$.

iv) The sequences $<y_n(s)>$, $<\bar{y}_n(s)>$ are also sequences of independent random variables.

v) Uniformly for $\alpha \in A$

$$\limsup_{N\to\infty} N^{-1} E\left\{\sum_{n\leqslant N} (z_n(\alpha) - E(z_n(\alpha)))\right\}^2 \overset{\text{def.}}{=} \sigma^2(\alpha) \leqslant 1$$

i.e. for any $\delta > 0$ there is an $M(\delta)$ independent of α such that $N^{-1}E\{\ldots\}^2 \leqslant \sigma^2(\alpha) + \delta$ for all $N \geqslant M(\delta)$.

Then for any $\varepsilon > 0$ there exists almost always in N_0 depending on ε and η such that

$$\frac{\sum_{n\leqslant N} \{z_n(\alpha) - E(z_n(\alpha))\}}{\sqrt{2N \log \log N}} < \sigma(\alpha) + \varepsilon$$

for all $\alpha \in A$ and $N \geqslant N_0$.

Before proving the theorem I would like to make several remarks. First, one can improve [14], theorem 1.3.2 in the same way and one can obtain similar improvements for families of mixing sequences of random variables.

Second, it is, perhaps, worthwhile to observe that the random variables $z_n(\alpha)$ are approximated by sums of $x_n(t)$ rather than by the $x_n(t)$'s themselves. A standard example is provided by the family $\{1_{[0,\alpha)}(Z_n(\omega))\}$ where the $Z_n(\omega)$ are independent identically distributed random variables, each having uniform distribution over $[0,1]$ and $1_{[0,\alpha)}(\cdot)$ denotes the indicator of the interval $[0,\alpha)$, $0 \leq \alpha \leq 1$. Next, expand α in a dyadic fraction $\alpha = \sum_{t=1}^{s} \varepsilon_t(\alpha)2^{-t} + \theta 2^{-s}$, $0 \leq \theta < 1$. We define $z_n(\alpha) = 1_{[0,\alpha)}(Z_n(\omega))$ and $\bar{y}_n(s) = 1_{[0,\beta_s)}(Z_n(\omega))$ with $\beta_s = \sum_{t \leq s} \varepsilon_t(\alpha)2^{-t}$. Similarly define $y_n(s)$. Finally, set $\bar{x}_n(t) = \bar{y}_n(t) - \bar{y}_n(t-1)$.

In later sections we shall deal with examples where the intervals $[0,\alpha)$ are replaced by rectangles, ellipsoids, convex sets etc.

The proof of Theorem 1 is similar but much less involved than for the mixing case considered in [14]. Let $\varepsilon > 0$ and $N \geq 1$ be given. Let T be the smallest integer A with

$$N^{1/2} \leq (\rho(A))^{1+\eta} \tag{8}$$

and let τ be the largest integer a with

$$\sum_{t=0}^{a} \rho(t) \leq \frac{1}{4}\, \varepsilon \log\log N . \tag{9}$$

We need two lemmas.

LEMMA 1. There exists an N_1 depending only on ε such that for all possible sequences $\langle y_n(\tau)\rangle$ and $N \geq N_1$ we have

$$P\left\{\max_{1\leq n\leq N} \sum_{j\leq n} \{y_j(\tau) - E(y_j(\tau)) > (1+\varepsilon)\{2s_N^2(y(\tau))\log\log N\}^{1/2}\right\} \leq (\log N)^{-1-\frac{1}{2}\varepsilon}.$$

Here

$$s_N^2(y(\tau)) = N\,\mathrm{Var}\{y_n(\tau)\}\ . \tag{10}$$

as the $y_n(\tau)$ are identically distributed.

Proof. The random variables $y_j(\tau) - E(y_j(\tau))$ are centered at expectation and are bounded by $2\tau \leq \log\log N$. By Minkowski's inequality, (ii) and (iii) it follows that

$$(\mathrm{Var}\{y_n(\tau)\})^{1/2} \leq \sum_{t=0}^{s} (\mathrm{Var}(x_n(t)))^{1/2} \ll \sum_{t=1}^{s} t^{-3} \ll 1.$$

The lemma follows from the usual argument used in the proof of the upper bound for the law of the iterated logarithm (see [10], pp. 261-262.) Actually, slightly modified this argument is used in the proof of

LEMMA 2. We have uniformly for all $\langle x_n(t)\rangle$ with $\tau \leq t \leq T$

$$P\left\{\max_{1\leq n\leq N} \sum_{j\leq n} \{x_j(t) - E(x_j(t))\} > (2t^{-3}N\log\log N)^{1/2}\right\} \leq \exp(-\rho(t)\log\log N).$$

Proof. From Lévy's maximal inequality ([10], p. 248, Remark) and (ii) we conclude that for $N \geq N_1$ the left hand side in lemma 2 does not exceed

$$2P\left\{\sum_{n\leq N}\{x_n(t) - E(x_n(t))\} > (t^{-3}N \log \log N)^{1/2}\right\}.$$

But now for any $w,k \geq 0$ and $N \geq N_1$

$$P\left\{\sum_{n\leq N}\{x_n(t)-E(x_n(t))\}\geq k\right\} \leq \exp(-wk)E\left\{\exp(w\sum_{n\leq N}\ldots)\right\}$$
$$\leq \exp(-wk)\prod_{n\leq N} E\left\{\exp\{w(x_n(t)-E(x_n(t)))\}\right\}$$
$$\leq \exp(-wk + \frac{1}{2} w^2 N(\rho(t))^{-1-\eta}(1+w))$$

by (10), (ii) and a well known inequality ([14], (1.3.16)). In the last inequality we put $w = (2t^3\rho^2(t)N^{-1}\log \log N)^{1/2}$ and $k = (t^{-3}N\log \log N)^{1/2}$ and obtain the result.

The remainder of the proof is exactly the same as in [14], pp. 28-29 except that the estimate (1.3.32) has to be replaced by

$$\prod_{t=0}^{\tau} \exp(\rho(t)) \cdot (\log N)^{-1-1/2\varepsilon} + \sum_{t=\tau+1}^{T} \exp(\rho(t))\exp(-\rho(t)\log \log N)$$
$$\leq (\log N)^{-1-1/4\varepsilon}$$

and that in the approximation argument for $\max \sum \{z_j(\alpha)-E(z_j(\alpha))\}$ we use (iii) and (8).

3. The Isotrope Discrepancy for r = 2

3.1. We shall give now two proofs that (7) holds for $r = 2$ and the class $\underline{S}$ of convex sets. For the first one we need the following lemma, due to Fejes Toth [4].

<u>LEMMA 3</u>. [4] Let C be a convex set contained in the unit square. Then for any integer $m \geq 3$ there exists an inscribed and a circumscribed m-gon whose areas differ by at most $20m^{-2}$.

Fejes Toth has a smaller constant, but this has no effect on the result to be proved. For $r = 3$ the order of approximation cannot be made smaller than m^{-1} as can be seen in the case of the sphere. This is where the argument breaks down for $r \geq 3$.

<u>LEMMA 4</u>. Let C be a convex set contained in the unit square. Then there exists a sequence P_t $(t = 2,3,4,\ldots)$ of inscribed and $Q_t(t = 2,3,4,\ldots)$ of circumscribed polygons with the following properties.

1) The number of vertices of P_t or Q_t does not exceed 2^t
2) $P_t \subset P_{t+1} \subset C \subset Q_{t+1} \subset Q_t$ for all $t = 2,3,4\ldots$
3) The area $\lambda(Q_t - P_t)$ of the difference of Q_t and P_t is

$$\lambda(Q_t - P_t) \leq 80.2^{-2t} .$$

For the proof use induction on t. Suppose that P_t and Q_t are already constructed. By Lemma 4 there are inscribed and circumscribed 2^t-gons whose areas differ by not more than 20.2^{-2t}. Let P_{t+1} be the polygon constructed by taking as vertices the vertices of P_t and of the inscribed 2^t-gon; construct Q_{t+1} similarly. They have the desired properties.

Let t be fixed. Partition each side of the unit square into 4^t equal parts. Subdivide the set of all possible polygons P_t and Q_t into classes in the following way: Extend all sides of P_t and record their points of intersection with the sides of the unit square. We put all those P_t into the same class for which these points of intersection fall into a given set of intervals of length 4^{-t} on the sides of the unit square. In this way we obtain at most $2^{(4t+4)2^t}$ classes. From each class we pick the largest and the smallest 2^t-gon. Their areas differ by not more than 4.4^{-t}.

We are now ready to apply Theorem 1. Let $\underline{C} = \{C(\alpha), \alpha \in A\}$ be the collection of all convex sets contained in the unit square and let $X_n(\omega)$ be a sequence of independent random vectors, each having distribution $P(X_n(\omega) \in B) = \lambda(B)$ for all Borel sets $B \subset U^2$ and $n \geq 1$. Here λ stands for the 2-dimensional Lebesgue-measure. Let $z_n(\alpha) = 1\{X_n(\omega) \in C(\alpha)\}$. For fixed $C(\alpha)$ and t determine P_t and Q_t according to Lemma 4. Find the smallest 2^t-gon R_t in the class of P_t and the largest 2^t-gon S_t in the class of Q_t. Then $R_t \subset P_t \subset C(\alpha) \subset Q_t \subset S_t$ and $\lambda(S_t - R_t) \leq 120.2^{-2t}$. The corresponding $y_n(t)$ and $\bar{y}_n(t)$ are defined as $y_n(t) = 1\{X_n(\omega) \in S_t\}$ and $\bar{y}_n(t) = 1\{x_n(\omega) \in R_t\}$.

We have $\bar{y}_n(t-1) \leqslant \bar{y}_n(t) \leqslant z_n(\alpha) \leqslant y_n(t) \leqslant y_n(t-1)$ for all t and n. Define $x_n(t) = y_n(t-1) - y_n(t)$ and $\bar{x}_n(t) = \bar{y}_n(t) - \bar{y}_n(t-1)$ and let A_t be the collection of all possible $\langle x_n(t)\rangle$ and $\langle \bar{x}_n(t)\rangle$. Then we can choose $\rho(t) \leqslant 10t.2^t$ and $\eta = \frac{1}{2}$.

3.2. We now turn to the second proof of the law of the iterated logarithm for the 2-dimensional isotrope discrepancies. Let $m \geqslant 1$ be integer and partition the unit cube U^r into m^r equal cubes of volume m^{-r} by dividing each edge of the cube into m equal parts.

Problem. Estimate the total number $\nu_r(m)$ of sets of such cubes through which the boundary of a convex set $C \subset U^r$ can pass through, no matter how C is chosen.

For $r = 1$ the convex sets are intervals, the boundary consists of the two endpoints. Thus the total number is less then m^2.

LEMMA 5. (W. Schmidt) We have $\nu_1(m) \leqslant m^2$ and $\nu_2(m) \ll \exp(c\, m^{2/3+\varepsilon})$ with $\varepsilon, c > 0$.

In order to apply Theorem 1 one would need the estimate $\nu_r(m) \ll \exp(m^{1-\delta})$ for some $\delta = \delta(r) > 0$. Actually, by refining the argument in the proof of Theorem 1 one even could get away with $\nu_r(m) \ll \exp(m \log^{-2} m)$.

Observing that any convex set contained in U^r cannot pass through more than Crm^{r-1} such cubes (for some positive constant C) one obtains

$$\nu_r(m) \leqslant \sum_{\ell \leqslant Crm^{r-1}} \binom{m^r}{\ell} \approx \binom{m^r}{Crm^{r-1}}$$

which turns out to be useless even for $r = 2$.

For the proof of lemma 5 we consider convex polygons P with vertices $(u/m, v/m)$ where $0 \leqslant u, v \leqslant m$ are integers. If $E_2(m)$ denotes the maximal number of vertices such a polygon can have then the total number $A_2(m)$ of such polygons does not exceed

$$A_2(m) \leqslant 1+(m+1)^2 + (m+1)^4 +\ldots+ (m+1)^{2E_2(m)}$$
$$\leqslant 2(m+1)^{2E_2(m)} .$$

For the estimate of $E_2(m)$ we subdivide the boundary of the convex set into at most eight parts in the following way: Along at most 2 pieces the slope varies between 0 and 1, along at most 2 more pieces between 1 and ∞, then between $-\infty$ and -1 and between -1 and 0. It is enough to estimate the number of vertices on one of these pieces, no matter how P is chosen and then raise the result to the eighth power. To fix the ideas let us consider the piece with slope between 0 and 1. The slopes of the sides of P are then of the form j/k with $0 \leqslant j \leqslant k$, $1 \leqslant k \leqslant m$ arranged in increasing order. Suppose that a_k of the sides have slope with denominator k. Then the inequalities $\Sigma a_k k \leqslant m$ and $a_k \leqslant k$ are satisfied. Our goal is then to estimate Σa_k subject to the above inequalities.

We split both sums into two parts Σ' and Σ'' where Σ' sums over those k's with $a_k < k^{1/2}$ and Σ'' over the remaining ones.* Denote the number of terms in Σ' and Σ'' by s and t respectively. Then $s << m^{1/2}$ and $t << m^{2/5}$. Hence by Hölder's inequality

$$\sum{}' a_k \leq \sum{}' (a_k k)^{1/3} \leq m^{1/3} s^{2/3} << m^{2/3}$$

and

$$\sum{}'' a_k \leq \sum{}'' (a_k k)^{1/2} \leq m^{1/2} s^{1/2} << m^{7/10}$$

By the remarks at the beginning of the proof the last two inequalities yield $A_2(m) << \exp(c\, m^{7/10})$ or in view of the footnote*) $A_2(m) << \exp(m^{2/3+\varepsilon})$.

By constructing polygons P which essentially pass through the same set of squares as the boundary of the given convex sets C it is easy to see that $\nu_2(m) << \exp(m^{2/3+\varepsilon})$.

We now apply Theorem 1. Let $\underline{C} = \{C(\alpha), \alpha \in A\}$ be again the collection of all convex sets contained in the unit square and let $X_n(\omega)$ be a sequence of independent random vectors, as defined in section 3.1. Define $z_n(\alpha) = 1\{X_n(\omega) \in C(\alpha)\}$. For fixed $C(\alpha)$ and t determine the set of squares as described in Lemma 5 with $m = 2^t$. Let R_t be the set of squares with side 2^{-t} completely covered by $C(\alpha)$ and $S_t = R_t \cup \{$set of squares through which $\partial C(\alpha)$ passes$\}$. Define $y_n(t), \bar{y}_n(t)$,

* This will only yield 7/10 instead of 2/3+ε in the exponent. But by splitting Σ'' further and treating the resulting sums in the same way one easily can get the desired exponent.

$x_n(t)$, and $\bar{x}_n(t)$ as in section 3.1. We can choose $\rho(t) = 2^{t(2/3+\varepsilon)}$ and $\eta = \frac{1}{3}$.

4. The Case of Ellipsoids

Let $S = \{S(\alpha), \alpha\ A\}$ be the class of ellipsoids with axex parallel to the coordinate planes. Consider all ellipsoids contained in $\underline{S}$ with center having coordinates of the form $k.2^{-t}$ $(k = 0,1,\ldots,2^t-1)$ and axes of length $j.2^{-t}$ $(j = 0,1,\ldots,2^t-1)$. There are 2^{2rt} such ellipsoids and every ellipsoid in $\underline{S}$ can be approximated by one containing it and one contained in it such that the difference of their volumina is $\ll 2^{-t}$. Denote these two ellipsoids by S_t and R_t so that $R_t \subset S(\alpha) \subset S_t$. Define $y_n(t)$, $\bar{y}_n(t)$, $x_n(t)$ and $\bar{x}_n(t)$ as in section 3.1 and apply Theorem 1. As a matter of fact, we even could apply [14], Theorem 1.3.1.

5. Sequences of the Form $\{A^n x\}$

Let A be a nonsingular $r \times r$ matrix and define the map $\underline{T}$ from U^r into U^r by $Tx = Ax \bmod 1$. Here $x = (x_1,\ldots,x_r)$ is an r-dimensional vector. The ergodic properties of maps of this type have been extensively studied by many authors, among them by Rohlin [16], Philipp [15] and Fischer [5]. For integral matrices Rohlin [16] gave necessary and sufficient conditions for A to define an exact endomorphism. For not necessarily integral matrices the author [15] gave an explicit representation for the measure μ invariant with respect to T provided that a certain function was integrable. If this function turns out to be essentially bounded then T is even an exact endomorphism and thus mixing of every degree. This is the case if, for example, all eigenvalues of A are greater than $r.2^{r+1}$ (see [15]). Under still more restrictive assumptions on the matrix A Fischer [5] proved a Gauß-Kusmin theorem. With the help of this theorem he proved that T is mixing with an exponential rate of decay for certain families of sets. We shall use Fischer's results to prove a law of the iterated logarithm for the discrepancies of the sequences $\{A^n x\}$.

We follow the presentation as given by Fischer [5]. For $x = (x_1,\ldots,x_r)$ write $[x] = ([x_1],\ldots,[x_r])$. Let $A.U^r = \langle Ax,\ x\in U^r\rangle$ and $M = AU^r \cap Z^r$, where Z denotes the set of integers. We define the "digits" $k_i(x)$ of a vector x inductively as maps from $U^r \to M$ given by

$$k_1(x) = [Ax],\quad k_i(x) = k_1(T^{i-1}x),\quad i = 2,3,\ldots$$

For $a_1,\ldots,a_s \in M$ the sets

$$I(a_1,\ldots,a_s) = \langle x\in U^r : k_i(x) = a_i,\quad 1 \leqslant i \leqslant s\rangle$$

are called cylinders of order s. They define a partition I_s of U^r. Let $A_s^* = \langle E\in I_s : T^s E = U^r\rangle$, $C_s^* = I_s - A_s^*$ and $D_s^* = \langle I(a_1,\ldots,a_s)\in I_s : I(a_1)\in C_1$, $I(a_1,a_2)\in C_2,\ldots,I(a_1,\ldots,a_s)\in C_s\rangle$. In other words A_s^* consists of those sets $E\in I_s$ which are mapped by T^s onto U^r, C_s^* of the "incomplete" ones and similarly D_s^*. We shall assume throughout that

$$(11)\qquad (\det A)^{-n} \operatorname{card} D_n^* \leqslant cq^n$$

for some constants c and $q < 1$. Under this hypothesis there exists a probability measure μ which is invariant with respect to T and equivalent to the Lebesque measure λ. We have

<u>LEMMA 6.</u> [5]: Let $I_s \in A_s^*$ and F be a Borel set. Then for all $n \geq s$

$$(12) \qquad |\lambda(I_s \cap T^{-n}F) - \lambda(I_s)\mu(F)| \leq c_1 \lambda(I_s)\lambda(F)\rho^{n-s}$$

and

$$(13) \qquad |\lambda(T^{-n}F) - \mu(F)| \leq c_1 \lambda(F)\rho^n$$

for some constants c_1 and $\rho < 1$, depending on A only.

As a matter of fact (12) and (13) are equivalent, i.e., each of them implies the other one (see [5], Lemma 1).

Next, for any Borel set E let $E^{(s)} = \bigcup H$, where the union is extended over all $H \subset A_\nu^* (\nu \leq s)$ with $H \subset E$. Put

$$(14) \qquad \psi_E(x) = \lambda(E - E^{(s)}) \quad \text{if} \quad s-1 < x \leq s .$$

With this notation we have

<u>LEMMA 7.</u> [5]: For any Borel sets E and F

$$|\lambda(E \cap T^{-n}F) - \lambda(E)\mu(F)| \leq c_2(\psi_E(\tfrac{n}{2}) + \rho^{n/2})$$

where c_2 depends on A only.

From Lemmas 6 and 7 we get easily the following

COROLLARY. Let E,F be Borel sets and let $I_s \in A_s^*$. Then for any $n \geq s$

$$|\lambda(I_s \cap T^{-n}F) - \lambda(I_s)\lambda(T^{-n}F)| \leq c_1 \lambda(I_s)\lambda(T^{-n}F)\rho^{n-s}$$

and

$$|\lambda(E \cap T^{-n}F) - \lambda(E)\lambda(T^{-n}F)| \leq c_2(\psi_E(\tfrac{n}{2}) + \rho^n)$$

where c_1, c_2 and $\rho < 1$ are constants depending on A only.

Let $\underline{S}$ be a class of Borel sets S contained in U^r. We suppose in addition to (11) that for all $S \in \underline{S}$

$$\psi_S(n) \leq c_3 q_1^n \tag{15}$$

where c_3 and $q_1 < 1$ are constants depending on the matrix A and on the class $\underline{S}$ only. It is easy to see that (15) holds in the following two cases with $\underline{S}$ the class of convex sets:

(i) $r = 1$ and $Tx = \beta x \bmod 1$ where $\beta > 1$. If $\beta = g$ is an integer then T yields the usual g-adic expansion of x. If β is not an integer then T describes Bolyai's algorithm.

(ii) A is a symmetric matrix with all eigenvalues greater than $r2^{r+1}$. A proof can easily be modeled after the proof of [15], Satz 3.

Let $A(N,S,x)$ be the number of indices $n \leq N$ with $A^n x \in S \bmod 1$. Define

$$D_N(\underline{S},x) = \sup_{S \in \underline{S}} |\tfrac{1}{N} A(N,S,x) - \lambda(S)|$$

THEOREM 2. Let A be a $r \times r$ matrix satisfying (11) and (15). Then for almost all $x \in U^r$

$$\limsup_{N\to\infty} \frac{ND_N(\underline{S},x)}{\sqrt{2N \log\log N}} = \max_{S \in \underline{S}} \sigma(S)$$

where

$$\sigma^2(S) = \lambda(S)(1-\lambda(S))+2\sum_{j=1}^{\infty}\{\int_{U^r} 1_S(x)1_S(T^j x)dx-\lambda^2(S)\}$$

and $\underline{S}$ is either the class of all rectangles contained in U^r or the class of ellipsoids contained in U^r with axes parallel to the coordinate planes.

The proof follows from the above corollary and [14], Theorem 1.3.1 in almost the same way as [14], Corollary 4.2.2. We restrict ourselves to the class $\underline{S}$ of ellipsoids with axes parallel to the coordinate planes. Let R_t and S_t be as in section 4 with $R_t \subset S(\alpha) \subset S_t$. Let $P_t = S_t^{(t)}$ and $\tilde{Q}_t = (\tilde{S}_t)^{(t)}$ where $\tilde{A}$ denotes the complement of A and $S_t^{(t)}$ is defined as in (14). From (15) we conclude that $\lambda(Q_t-P_t) \leq 2c_3q_1^t$. Define $z_n(\alpha) = 1\{A^n x \in S(\alpha) \bmod 1\}$, $y_n(t) = 1\{A^n x \in Q_t \bmod 1\}$, $\bar{y}_n(t) = 1\{A^n x \in P_t \bmod 1\}$ and $x_n(t) = y_n(t-1)-y_n(t)$, $\bar{x}_n(t) = \bar{y}_n(t)-\bar{y}_n(t-1)$. Let A_t be the collection of all $\langle x_n(t)\rangle$ and $\langle \bar{x}_n(t)\rangle$ which can be obtained in this way if $S(\alpha)$ ranges through $\underline{S}$. As in section 4 we have card $A_t \leq 2^{2rt}$. The corollary and [14], (1.3.2) are satisfied. Also [14], (1.3.4) follows from the corollary. The remaining conditions of [14], Theorem 1.3.1 are checked in a similar fashion.

Finally we observe that for any strictly increasing sequence $<\ell_n>$ of positive integers we have

<u>THEOREM 3</u>. Let A be an $r \times r$ matrix satisfying (11) and (15) and let $D_N(\underline{S},x)$ be the rectangular or elliptic discripancy for the sequence $<A^{\ell_n}x>$ mod 1 as defined above. Then for almost all $x \in U^r$

$$\limsup_{N\to\infty} \frac{ND_N(\underline{S},x)}{\sqrt{2 N \log\log N}} \leq C$$

where C is a constant depending on the matrix A only.

The proof follows in the similar way as [14], Corollary 4.2.3.

Note added later: The following lemma shows that the bounds for $\nu_r(m)$, $r \geq 3$ are too large to yield the law of the iterated logarithm for the isotrope discrepancy in dimension $r \geq 3$.

<u>LEMMA</u>:

$$\nu_3(m) >> 2^{1/8\, m}$$

For the proof consider the sphere S of radius 1/4 with center (1/2, 1/2, 1/2). Let P be a point of distance $2m^{-1}$ from S. The tangential cone C(P) generated by P covers a spherical cap S(P) whose area has order of magnitude m^{-1}. Hence there are at least cm points P_n with nonoverlapping caps $S(P_n)$. Elementary geometry shows $c \geq 1/8$. For any choice of points $P_{n_1}, \ldots, P_{n_k}$, $k \leq \frac{1}{8} m$ the convex hull of $P_{n_1}, \ldots, P_{n_k}$ and S is the union $S \cup \bigcup_{j \leq k} S(P_{n_j})$. We can select these points in $2^{1/8 m}$ different ways and each selection yields a different convex set. Since the distance of P_n from S is bigger than the diagonal of a cube with side m^{-1} each of these convex sets defines a different system of covering cubes.

Acknowledgement: I am grateful to Professor W. Schmidt for pointing out that Lemma 5 in the original manuscript was false and for providing the present version of the lemma and its proof.

REFERENCES

[1] Patrick Billingsley, Convergence of probability measures, Wiley,1968.

[2] J.W.S. Cassels, An extension of the law of the iterated logarithm, Proc. Cambridge Philos. Soc. 47,(1951), 55-64.

[3] K.L. Chung, An estimate concerning the Kolmogoroff limit distribution, Trans. A.M.S. 67, (1949), 36-50.

[4] L. Fejes Toth, Über die Approximation konvexer Kurven durch Polygonfolgen, Compositio Math. 6,(1938-39), 456-467.
Eine Bemerking zur Approximation durch n-Eckringe Compositio Math 7 (1939-40), 474-476.

[5] Roland Fischer, Mischungsgeschwindigkeit für Ziffernentwicklungen nach reellen Matrizen, to appear.

[6] S. and L. Gaal, The discrepancy of the sequence $\{(2^n x)\}$, Proc. Koningl. Nederl. Akad. Wetensch. Ser. A 67 (1964), 129-134.
[7] J. Kiefer, On large deviations of the empiric d.f. of vector chance variables and a law of the iterated logarithm, Pacific J. Math. 11, (1961), 649-660.
[8] J. Kiefer, and J. Wolfowitz, On the deviations of the empiric distribution function of vector chance variables, Trans. A.M.S., 87, (1958), 173-186.
[9] A.N. Kolmogorov, Sulla determinazione empirica di una legge di distribuzione, Inst. Ital. Atti, Gion. 4, (1933), 83-91.
[10] Michel Loéve, Probability theory, 3rd ed., Van Nostrand, Princeton, 1963.
[11] H. Niederreiter, Metric theorems on the distribution of sequences, Proc. Sympo. Pure Math. vol. 24, AMS, Providence, R.I., 1973.
[12] H. Niederreiter, oral communication.
[13] Walter Philipp, Das Gesetz vom iterierten Logarithmus mit Anwendungen auf die Zahlentheorie, Math. Annalen 180, (1969), 75-94, Corrigen dum, Ibid., 190, (1971), 338.
[14] Walter Philipp, Mixing sequences of random variables and probabilistic number theory, Memoirs A.M.S. vol 114, (1971).
[15] Walter Philipp, Mischungseigenschaften gewisser auf dem Torus definierter Endomorphismen, Math. Zeitschr. 101, (1967), 369-374.
[16] V.A. Rohlin, Exact endomorphisms of a Lebesgue space, Isvest. Akad. Nauk, 25 (1961), 499-530; AMS Transl. Ser. 2, 39, (1964), 1-36.
[17] Wolfgang M. Schmidt, Irregularities of distribution IV., Inventiones math. 7, (1969), 55-82.
[18] Olaf P. Stackelberg, A uniform law of the iterated logarithm for functions C-uniformly distributed mod 1, Indiana Math. J. 21 (1971), 515-528.
[19] S.K. Zaremba, Sur la discrépance des suites aléatoires, Z. Wahrscheinl. verw. Geb. 20 (1971), 236-248.

[20] S.K. Zaremba, La discrépance isotrope et l'integration numeŕique, Annali di Mat 87 (1970), 125-136.

University of Illinois, Urbana, Illinois

INEQUALITIES FOR RESULTANTS AND FOR DECOMPOSABLE FORMS

Wolfgang M. Schmidt

1. Resultants

We shall deal with resultants of polynomials with integer coefficients. The resultant of polynomials P, Q will be denoted by $r(P,Q)$. The *height* of a polynomial P, i.e. the maximum of the absolute values of its coefficients, will be denoted by $H(P)$.

THEOREM 1. Suppose $P(x)$ is a polynomial of degree d, with integer coefficients and without multiple roots. Suppose $0 < t < d$, and $P(x)$ has no factor of degree $\leqq t$ with integer coefficients. Then for every $\varepsilon > 0$, there are only finitely many polynomials $Q(x)$ of degree $\leqq t$ with integer coefficients and with

$$(1) \qquad |r(P,Q)| < H(Q)^{d-2t-\varepsilon}.$$

A theorem of this type had been proved by Wirsing [6, Theorem 5]. He needed the more stringent inequality

$$0 < |r(P,Q)| < H(Q)^{d-\Phi-\varepsilon}$$

with $\Phi = 2t(1 + (1/3)+\ldots+(1/(2t-1)))$ instead of (1), but he did not require that $P(x)$ has no factor of degree $\leqq t$. It is easily seen that the case $t = 1$ of Theorem 1 is a consequence of Roth's Theorem on approximation to algebraic numbers. Fujiwara [2] showed (but did not express his result in this way) that if $P(x)$ is irreducible and if $d > 2t$, then there are only finitely many polynomials $Q(x)$ of degree $\leqq t$ with $r(P, Q) = c$, where c is a constant.

We now shall discuss in how far the exponent in (1) is best possible. If

$$P(x) = a_d x^d+\cdots+a_0 = a_d(x-\alpha^{(1)})\cdots(x-\alpha^{(d)}) ,$$

we have

$$r(P,Q) = a_d^q Q(\alpha^{(1)})\cdots Q(\alpha^{(d)}) , \tag{2}$$

where q is the degree of Q. If one of the roots $\alpha^{(1)},\ldots,\alpha^{(d)}$ is real, say $\alpha^{(1)} = \alpha$ is real, then Dirichlet's Theorem on approximation to linear forms implies that for every large H there is a non-zero polynomial Q of degree $\leqq t$ with

$$|Q(\alpha)| < c_1 H^{-t} \quad \text{and} \quad H(Q) \leqq H ,$$

where $c_1 = c_1(\alpha)$. Since α, being a root of P, is of degree greater than t, we have $Q(\alpha) \neq 0$. Hence there are infinitely many polynomials Q of degree $\leqq t$ with

$$|Q(\alpha)| < c_1 H(Q)^{-t} . \tag{3}$$

If, on the other hand, $\alpha^{(1)}, \alpha^{(2)}$ are complex conjugate roots of P, and if $t \geqq 2$, one sees in a similar way that there are infinitely many polynomials Q of degree $\leqq t$ with

$$(4) \qquad |Q(\alpha^{(1)})| = |Q(\alpha^{(2)})| < c_2 H(Q)^{-(t-1)/2} .$$

By virtue of (2), and since $|Q(\alpha^{(i)})| \leqq c_3 H(Q)$ $(i = 1,\ldots,d)$, each of the inequalities (3), (4) implies that

$$(5) \qquad |r(P,Q)| < c_4 H(Q)^{d-t-1} .$$

Hence the exponent $d-2t-\varepsilon$ in (1) may not be replaced by a number greater than $d-t-1$.

In fact, there are cases when $d-2t$ in the exponent in (1) is best possible. Suppose $d = st$ where s is an integer greater than 1. Let $P_0(y)$ be an irreducible polynomial of degree s with a real root θ. According to (5), there are infinitely many polynomials $Q_0(y) = b_1 y + b_0$ of degree 1 with

$$(6) \qquad 0 < |r(P_0,Q_0)| < c_4 H(Q_0)^{s-2} .$$

Now suppose that the polynomial $x^t - \theta$ is irreducible over the field $\underline{Q}(\theta)$ obtained by adjoining θ to the field $\underline{Q}$ of rationals. Then $P(x) = P_0(x^t)$ is of degree $d = st$ and is irreducible over the rationals. Given a polynomial $Q_0(y)$ with (6), put $Q(x) = Q_0(x^t)$. If α runs through the $d = st$ roots $\alpha^{(1)},\ldots,\alpha^{(d)}$ of P, then α^t runs t times through the s roots $\theta^{(1)},\ldots,\theta^{(s)}$ of P_0. Hence

$$\begin{aligned} r(P,Q) &= a_d^t Q(\alpha^{(1)})\ldots Q(\alpha^{(d)}) \\ &= (a_d Q_0(\theta^{(1)})\ldots Q_0(\theta^{(s)}))^t \\ &= (r(P_0,Q_0))^t , \end{aligned}$$

and

$$0 < |r(P,Q)| < c_5 H(Q_0)^{st-2t} = c_5 H(Q)^{d-2t} .$$

We finally remark that the inequality (1) of Theorem 1 may be replaced by

$$(7) \qquad |r(P,Q)| < H(Q)^{d-t-1-\varepsilon} ,$$

if one makes stronger assumptions on P. It follows from Theorem 3 in [5] that we may use (7) instead of (1) if P is irreducible and if the Galois group of the splitting field of P is t times transitive on the roots of P.

2. Decomposable Forms

A form $f(\underline{\underline{x}}) = f(x_1,\ldots,x_n)$ is called *decomposable* if it is a product of linear forms with algebraic coefficients. The *rank* of a set of linear forms $\ell_1(\underline{\underline{x}}),\ldots,\ell_h(\underline{\underline{x}})$ is the maximum number of linearly independent forms among them.

THEOREM 2. Suppose $f(\underline{\underline{x}})$ is a decomposable form of degree d with rational coefficients. Suppose $1 < n \leq k < d$ and $f(\underline{\underline{x}})$ is not divisible by a rational form of degree less than k, and suppose any k linear factors $\ell_i(\underline{\underline{x}})$ in the product $f(\underline{\underline{x}}) = \ell_1(\underline{\underline{x}})\ldots\ell_d(\underline{\underline{x}})$ have rank n. Then for every

$\varepsilon > 0$, there are only finitely many integer points $\underline{\underline{x}}$ with

$$(8) \qquad |f(\underline{\underline{x}})| < |\underline{\underline{x}}|^{d-2(k-1)-\varepsilon},$$

where $|\underline{\underline{x}}| = \max(|x_1|,\ldots,|x_n|)$.

When $n = k = 2$, (8) becomes $|f(\underline{\underline{x}})| < |\underline{\underline{x}}|^{d-2-\varepsilon}$, and we have the same exponent $d-2-\varepsilon$ which is obtained if Roth's Theorem is applied in this context ([3]).

Theorem 1 follows from the case $n = k$ of Theorem 2. To see this, we note that by (2), $r(P,Q) = a_d^q\, r^*(P,Q)$ with

$$r^*(P,Q) = Q(\alpha^{(1)})\ldots Q(\alpha^{(d)})$$

$$= \ell_1(\underline{\underline{b}})\ldots\ell_d(\underline{\underline{b}}) ,$$

where $Q(x) = b_t x^t+\ldots+b_1 x+b_0$ and where $\ell_i(\underline{\underline{b}}) = \ell_i(b_t,\ldots,b_1,b_0)$ is the linear form

$$\ell_i(\underline{\underline{b}}) = b_t(\alpha^{(i)})^t+\ldots+b_1\alpha^{(i)}+b_0$$

in the $n = t+1$ variables $b_t,\ldots,b_1,b_0$. Thus $r^*(P,Q)$ is a decomposable form $f(\underline{\underline{b}})$ in $\underline{\underline{b}}$ of degree d. If $P(x) = P_1(x)\ldots P_h(x)$ is the factorization of P into polynomials with rational coefficients which are irreducible over the rationals, and if $g_j(\underline{\underline{b}})$ is the product of the linear forms $\ell_i(\underline{\underline{b}})$ with subscripts i satisfying $P_j(\alpha^{(i)})= 0$, then $f(\underline{\underline{b}}) = g_1(\underline{\underline{b}})\ldots g_h(\underline{\underline{b}})$. The forms g_j clearly have rational coefficients. We claim that they are irreducible over the rationals. For suppose to the contrary that, say, $g_1(\underline{\underline{b}}) = h_1(\underline{\underline{b}})$ $h_2(\underline{\underline{b}})$ with $0 < \deg h_1 < \deg g_1 = \deg P_1$. If $h_1(\underline{\underline{b}}) = \ell_{i_1}(\underline{\underline{b}})\cdots\ell_{i_r}(\underline{\underline{b}})$, then the coefficients of the monomials $b_1^u b_0^{r-u}$ $(1 \leq u \leq r)$ in $h_1(\underline{\underline{b}})$ are the elementary symmetric functions in $\alpha^{(i_1)},\ldots,$ $\alpha^{(i_r)}$. Thus the numbers $\alpha^{(i_1)},\ldots,\alpha^{(i_r)}$ are algebraic of degree $\leq r = \deg h_1 < \deg P_1$, which contradicts the fact that they are roots of the irreducible polynomial P_1.

Now by the hypothesis of Theorem 1, P has no factor of degree $\leq t$ with integer coefficients, hence by Gauss' Lemma, no such factor with rational coefficients. Hence each irreducible factor $g_j(\underline{\underline{b}})$ of $f(\underline{\underline{b}})$ has a degree $\geq t + 1 = n$, and $f(\underline{\underline{b}})$ has no rational factor of degree less than n. By using Van der Monde determinants, one sees that any $n = t+1$ linear forms ℓ_i have rank n.

Thus Theorem 2 applies with $k = n$, and there are only finitely many integer points $\underline{\underline{b}}$ with $|f(\underline{\underline{b}})| < |\underline{\underline{b}}|^{d-2(n-1)-\varepsilon}$, hence only finitely many polynomials Q of degree $\leq t$ with (1).

Sometimes it is desirable to have a result which does not depend on a condition on the ranks of the linear forms. We shall restrict ourselves to the case when $f(\underline{\underline{x}})$ is decomposable, but is irreducible over the rationals. In this case it is well known that $f(\underline{\underline{x}})$ is a constant times a norm form (see, e.g., [1, chapter 2] or [4]).

A norm form is a form $\mathfrak{N}(\alpha_1 x_1+\ldots+\alpha_n x_n)$, where $\alpha_1,\ldots,\alpha_n$ are fixed elements of an algebraic number field K, and where $\mathfrak{N}$ denotes the norm from K to the rationals. Such a norm form is irreducible precisely if K is generated by the quotients α_i/α_j with $\alpha_j \neq 0$.

THEOREM 3. Suppose $N(\underline{\underline{x}}) = \mathfrak{N}(\alpha_1 x_1+\ldots+\alpha_n x_n)$ is an irreducible norm form of degree $d > n > 1$, with coefficients $\alpha_1,\ldots,\alpha_n$ which are linearly independent over the rationals and lie in a number field K which is primitive (i.e. it has no non-trivial subfields). Then for every $\varepsilon > 0$, there are only finitely many integer points $\underline{\underline{x}}$ with

$$|N(\underline{\underline{x}})| < |\underline{\underline{x}}|^{d-d'-\varepsilon} , \tag{9}$$

where

$$d' = \frac{n}{n-1}\, 2^{(n-2)/2}\, d^{(n-2)/(n-1)} . \tag{10}$$

When $n = 2$ we have $d' = 2$; but of course this case follows from Roth's Theorem. The first new case is when $n = 3$, with $d' = (3/\sqrt{2})\sqrt{d}$.

3. Proof of Theorem 2.

The following is an immediate consequence of Theorem 1 in [4].

THEOREM A. Suppose $f(\underline{\underline{x}}) = \ell_1(\underline{\underline{x}})\ldots\ell_d(\underline{\underline{x}})$ has rational coefficients and is a product of linear forms with real or complex algebraic coefficients. If the inequality

(11) $$|f(\underline{\underline{x}})| < c|\underline{\underline{x}}|^{d-\eta}$$

has infinitely many integer solutions $\underline{\underline{x}}$ for some $c > 0$, $\eta > 0$, then there is a rational subspace R^t of dimension t with $1 \leqq t \leqq n$, and there are m linear forms $\ell_{i_1},\ldots,\ell_{i_m}$ with $1 \leqq m \leqq d$ and with distinct $i_1,\ldots,i_m$, whose restrictions to R^t have a rank r with

$$r \leqq tm/\eta \quad \text{and} \quad r < t.$$

We have to show that under the hypotheses of Theorem 2, (11) cannot hold with $\eta > 2(k-1)$. In view of Theorem A, it will be enough to verify that we cannot have $r < tm/(2(k-1))$ and $r < t$, i.e. to show that

(12) $$r < t \quad \text{implies} \quad tm \leqq 2(k-1)r .$$

If $\ell(\underline{\underline{x}})$ is a linear form with algebraic coefficients and with its first non-zero coefficient equal to 1, write $F(\ell)$ for the number field generated by the coefficients of ℓ. We shall say that two such forms $\ell(\underline{\underline{x}})$, $\ell'(\underline{\underline{x}})$ are <u>conjugate</u>, if there is an isomorphism from $F(\ell)$ onto $F(\ell')$ which maps the coefficients of $\ell(\underline{\underline{x}})$ into the respective coefficients of $\ell'(\underline{\underline{x}})$. A form $\ell(\underline{\underline{x}})$ has precisely f conjugates where f is the degree of $F(\ell)$, and no two conjugates are proportional.

In proving Theorem 2, we may assume without loss of generality that all the linear forms ℓ_i in $f(\underline{\underline{x}}) = \ell_1(\underline{\underline{x}})\ldots\ell_d(\underline{\underline{x}})$ have their first nonzero coefficient equal to 1. With a form $\ell_i(\underline{\underline{x}})$, also all its conjugates divide $f(\underline{\underline{x}})$. If we block together conjugate linear forms, we may write

$$f(\underline{\underline{x}}) = f_1(\underline{\underline{x}}) \dots f_\ell(\underline{\underline{x}}) ,$$

where each f_j is the product of conjugate linear forms. Each form f_j, being a product of conjugate linear forms, has rational coefficients, and hence has degree $\geqq k$ by the hypothesis of Theorem 2. Hence every linear form $\ell_i(\underline{\underline{x}})$ has at least k conjugates.

Now if $\ell_i(\underline{\underline{x}}) = 0$ for some integer point $\underline{\underline{x}}$, then also $\ell_{i_j}(\underline{\underline{x}}) = 0$ for the conjugates $\ell_{i_1} = \ell_i, \ell_{i_2}, \dots, \ell_{i_f}$ of ℓ_i. We have $f \geqq k$, so that by our hypothesis the forms $\ell_{i_1}, \dots, \ell_{i_f}$ have rank n, and $\underline{\underline{x}}$ must be $\underline{\underline{0}}$. Therefore $\ell_i(\underline{\underline{x}}) \neq 0$ $(1 \leqq i \leqq d)$ for every integer point $\underline{\underline{x}} \neq \underline{\underline{0}}$. Hence we have rank $r > 0$ on every rational subspace R^t of dimension $t \geqq 1$. We therefore have to show (12) only for $r > 0$, i.e. we have to show that

(13) $\quad 0 < r < t$ implies $tm \leqq 2(k-1)r$.

Let R^t be an arbitrary fixed rational subspace of R^n of dimension $t \geqq 2$; we shall prove (13) for R^t. Let $m(r)$ be the maximum number of linear forms $\ell_{i_1}, \dots, \ell_{i_m}$ whose restrictions to R^t have rank r, and put $q(r) = m(r)/r$. We have to show that

$$q(r) \leqq 2(k-1)/t \qquad (0 < r < t) .$$

Let q_0 be the maximum value of $q(r)$ in $0 < r < t$. Let r_0 be the largest number in $0 < r < t$ with $q(r) = q_0$. Put $m = m(r_0)$; there exists an m-tuple $\ell_{i_1},\ldots,\ell_{i_m}$ of linear forms whose restrictions to R^t have rank r_0, hence rank $< t$. It follows that the forms ℓ_{i_1}, $\ldots,\ell_{i_m}$ have rank $< n$ on the full n-dimensional space R^n. By the hypothesis of Theorem 2, we may conclude that

(14) $$m < k .$$

Now if ℓ_i occurs in an m-tuple of linear forms whose restrictions to R^t have rank r_0, then so does every conjugate of ℓ_i. Since ℓ_i has at least k conjugates and since $m < k$, there must be two distinct m-tuples $\ell_{i_1},\ldots,\ell_{i_m}$ and $\ell_{j_1},\ldots,\ell_{j_m}$ whose restrictions to R^t have rank r_0. Let the intersection of these two m-tuples consist of m^- linear forms with rank r^- on R^t, and let the union of these two m-tuples consist of m^+ linear forms with rank r^+ on R^t. Then $m^+ > m$,

(15) $$m^- + m^+ = 2m \quad \text{and} \quad r^- + r^+ \leqq 2r_0 .$$

Now if $m^- = 0$, we have $r^- = 0$ and $q(r^+) \geqq m^+/r^+ \geqq m/r_0 = q_0$. If $m^- > 0$, then $0 < r^- \leqq r_0$ and $m^-/r^- \leqq q(r^-) \leqq q(r_0) = m/r_0$, whence again $q(r^+) \geqq m^+/r^+ \geqq q_0$. Now clearly $r^+ \geqq r_0$, and if we had $r^+ = r_0$, then $q(r^+) = q_0 = m/r_0$ and $m^+ = m$, which is impossible. Hence $r^+ > r_0$, and since r_0 was maximal in $0 < r < t$ with $q(r) = q_0$, we must have $r^+ = t$. Now (14), (15) yield $t \leqq 2r_0$ and

$$q_0 = m/r_0 \leqq (k-1)/r_0 \leqq 2(k-1)/t \ .$$

4. Lemmas about Vectors.

Let V be a vector space. Given a finite set A of vectors, we write $z(A)$ for the number of its elements, $\langle A\rangle$ for the subspace spanned by A, and $\dim A$ for the dimension of $\langle A\rangle$. A 1-1 mapping ϕ from a finite set S onto a finite set T will be called an <u>isomorphism</u> if $\dim\phi(A) = \dim A$ for every subset A of S. If ϕ is such an isomorphism and if A,B are subsets of S with

$$\langle A\rangle \subseteqq \langle B\rangle, \quad \text{then} \quad \langle\phi(A)\rangle \subseteqq \langle\phi(B)\rangle \ .$$

For $\langle A\rangle \subseteqq \langle B\rangle$ implies that $\langle B\rangle = \langle A \cup B\rangle$, whence $\dim B = \dim(A \cup B)$, whence $\dim\phi(B) = \dim(\phi(A \cup B)) = \dim(\phi(A) \cup \phi(B))$, whence $\langle\phi(B)\rangle = \langle\phi(A) \cup \phi(B)\rangle$, whence $\langle\phi(A)\rangle \subseteqq \langle\phi(B)\rangle$.

The isomorphisms ϕ from a finite set S onto itself form a group $G(S)$ under composition. The set S will be called <u>regular</u>, if for every subset A of S with $1 < z(A) < z(S)$, there is a $\phi \in G(S)$ with $\phi(A) \neq A$ and $\phi(A) \cap A \neq \emptyset$.

LEMMA 1. Suppose S is a regular set with $z(S) \geq 3$. Then $G(S)$ is transitive on S, i.e. given vectors $\underline{\underline{a}}, \underline{\underline{b}}$, in S, there is a $\psi \in G(S)$ with $\psi(\underline{\underline{a}}) = \underline{\underline{b}}$.

Proof. Let $A(\underline{\underline{a}})$ be the set of vectors $\underline{\underline{b}} = \psi(\underline{\underline{a}})$ with $\psi \in G(S)$, and let $B(\underline{\underline{a}})$ be the complement of $A(\underline{\underline{a}})$ is S. Then $\phi(A(\underline{\underline{a}})) = A(\underline{\underline{a}})$ and $\phi(B(\underline{\underline{a}})) = B(\underline{\underline{a}})$ for every $\phi \in G(S)$. Hence $z(A(\underline{\underline{a}}))$ equals $0,1$ or $z(S)$, and also $z(B(\underline{\underline{a}}))$ equals $0,1$ or $z(S)$. Now $z(A(\underline{\underline{a}}))$ cannot be 0 since $\underline{\underline{a}} \in A(\underline{\underline{a}})$, and $z(A(\underline{\underline{a}}))$ cannot be 1 since $z(A(\underline{\underline{a}})) + z(B(\underline{\underline{a}})) = z(S) \geq 3$. Therefore $z(A(\underline{\underline{a}})) = z(S)$ and $A(\underline{\underline{a}}) = S$.

In what follows, S will be a fixed regular set with $z(S) \geq 3$. The number r will be a fixed integer with

$$(16) \qquad 1 < r < \dim S.$$

Let m be the maximum value of $z(A)$ for all subsets $A \subseteq S$ with $\dim A = r$. A subset A with $\dim A = r$ and $z(A) = m$ will be called special. An isomorphism ϕ of S carries special subsets into special subsets. Since $G(S)$ is transitive on S, there is a number $k = k(S,r)$ such that every vector $\underline{\underline{a}}$ of S lies in precisely k special sets. In view of (16), we have $1 < m = z(A) < z(S)$ for a special set A, and hence there is a $\phi \in G(S)$ with $\phi(A) \neq A$ and $\phi(A) \cap A \neq \emptyset$. Every vector $\underline{\underline{a}}$ in $\phi(A) \cap A$ lies in the two special sets A and $\phi(A)$, and hence we must have

$$k \geqq 2 .$$

Given a subset B of S, write $\ell(B)$ for the number of special sets A with $\langle B \rangle \subseteqq \langle A \rangle$. Write ℓ_u for the maximum of $\ell(B)$ over all B with $\dim B = u$. Then $\ell_1 \geqq \ldots \geqq \ell_r$.

<u>LEMMA 2.</u> $\quad k - \ell_u \geqq 2^{u-r}(k-1) \qquad (2 \leqq u \leqq r)$.

<u>Proof</u>. We begin with the case $u = r$. We have to show in this case that $\ell_r \leqq 1$. Now if $\dim B = r$ and if $\langle B \rangle \subseteqq \langle A \rangle$ with a special set A, then $\langle B \rangle = \langle A \rangle$. There can be at most one special set with this property, for if we had two, say A and A′, then $\dim(A \cup A') = r$ and $z(A \cup A') > m$, contradicting our choice of m.

To finish the proof of the lemma, it will suffice to show that

$$k - \ell_u \geqq \frac{1}{2}(k - \ell_{u+1}) \qquad (2 \leqq u < r). \tag{17}$$

So let B be a subset of S with $\dim B = u$ where $2 \leqq u < r$; we shall deduce the inequality $k - \ell(B) \geqq \frac{1}{2}(k - \ell_{u+1})$. We may assume that B is maximal in the sense that $B \subseteqq B'$, $B \neq B'$ implies $\langle B \rangle \neq \langle B' \rangle$. Suppose the $\ell = \ell(B)$ special sets $A_1, \ldots, A_\ell$ have $\langle B \rangle \subseteqq \langle A_i \rangle$ $(i = 1, \ldots, \ell)$. Since $2 \leqq u = \dim B < r < \dim S$, we have $2 \leqq z(B) < z(S)$. Hence there is a $\phi \in G(S)$ with $\phi(B) \neq B$ and $\phi(B) \cap B \neq \emptyset$. We have $\langle \phi(B) \rangle \subseteqq \langle \phi(A_i) \rangle$ $(i = 1, \ldots, \ell)$. Now B is a proper subset of $B \cup \phi(B)$, and since B was maximal, we have $\dim(B \cup \phi(B)) \geqq (\dim B) + 1 = u + 1$. Hence at most ℓ_{u+1} special sets A have $\langle B \cup \phi(B) \rangle \subseteqq \langle A \rangle$, i.e. have $\langle B \rangle \subseteqq \langle A \rangle$ and $\langle \phi(B) \rangle \subseteqq \langle A \rangle$. Hence there must be (at least) $2\ell - \ell_{u+1}$ distinct special sets among $A_1, \ldots, A_\ell$, $\phi(A_1), \ldots, \phi(A_\ell)$. These $2\ell - \ell_{u+1}$ special sets A all satisfy $\langle B \cap \phi(B) \rangle \subseteqq \langle A \rangle$, and since special sets were chosen maximal with rank r, they satisfy $B \cap \phi(B) \subseteqq A$. Thus every vector in $B \cap \phi(B)$ is contained in at least $2\ell - \ell_{u+1}$ special sets, and we have $2\ell - \ell_{u+1} \leqq k$, whence

$$k - \ell(B) = k - \ell = \tfrac{1}{2}(k + (k - 2\ell)) \geqq \tfrac{1}{2}(k - \ell_{u+1}) .$$

Since B was arbitrary with $\dim B = u$, (17) follows.

LEMMA 3. Suppose S is a regular set with $z(S) \geqq 3$. For r with $0 \leqq r < \dim S$, write $m_r = m_r(S)$ for the maximum of $z(B)$ over all subsets $B \subseteqq S$ with $\dim B = r$. Then

$$m_0 = 0 ,$$

$$m_r \leqq 2^{(r-1)/2} z(S)^{(r-1)/r} \qquad (1 \leqq r < \dim S) .$$

Proof. If $B \subseteqq S$ has $\dim B = 0$, then also $\dim \phi(B) = 0$ for $\phi \in G(S)$, and since $G(S)$ is transitive on S, we have either $z(B) = 0$ or

$\dim S = 0$. Since by hypothesis $\dim S > 0$, we obtain $z(B) = 0$. This shows that $m_0 = 0$.

Now suppose $1 < \dim S$, and B is a subset of S with $\dim B = 1$, $z(B) = m_1$. If we had $1 < m_1$, then $1 < m_1 = z(B) < z(S)$, and there is a $\phi \in G(S)$ with $\phi(B) \neq B$, $\phi(B) \cap B \neq \emptyset$. Then $\dim(B \cup \phi(B)) = 1$ and $z(B \cup \phi(B)) > m_1$, contradicting our choice of m_1. Therefore $m_1 \leqq 1$, and the lemma is true for m_1. To finish the proof, we may assume that $1 < r \leqq \dim S$, i.e. that (16) holds. We shall now keep r fixed, and we shall talk again of special sets and use the notations $m = m_r$ and $k = k(S,r)$.

Now let $1 < u \leqq r$, and let B be a set with $\dim B = u$, $z(B) = m_u$. For each $\underline{a} \in B$, there are k special sets containing $\underline{a}$, and $\geqq k-\ell_u$ special sets A with $\underline{a} \in A$ and $\langle B\rangle \not\subseteq \langle A\rangle$. For such a special set A, $\dim(B \cap A) \leqq \dim(\langle B\rangle \cap \langle A\rangle) < \dim B$, whence $\dim(B \cap A) \leqq u-1$. This in turn implies that $z(B \cap A) \leqq m_{u-1}$. Since $z(B) = m_u$, there must be

$$\geqq (m_u/m_{u-1})(k-\ell_u)$$

special sets A with $A \cap B \neq \emptyset$ and $\langle B\rangle \not\subseteq \langle A\rangle$. On using Lemma 2, we see in particular that there are

$$\geqq (m_u/m_{u-1})2^{u-r}(k-1)$$

special sets. In each special set are $m = m_r$ vectors, and since each vector lies in k special sets, the total number $z(S)$ of vectors in S satisfies

$$z(S) \geqq (m/k)(m_u/m_{u-1})2^{u-r}(k-1)$$

$$\geqq m(m_u/m_{u-1})2^{u-r-1} .$$

If we apply this inequality for $u = 2,\ldots,r$, we obtain

$$z(S)^{r-1} \geqq m^{r-1}(m_r/m_1)2^{-r(r-1)/2} .$$

Since $m_r = m$ and since, as we have shown at the beginning, $m_1 = 1$, we have $z(S)^{r-1} \geqq m_r^r 2^{-r(r-1)/2}$, and the lemma follows.

It may be shown that the exponent $(r-1)/r$ of $z(S)$ in Lemma 3 is best possible if $0 < r = \dim S - 1$. Suppose $n > 1$, and for large t, choose algebraically independent numbers $x_{21},\ldots,x_{2t}$, $x_{31},\ldots,x_{3t}$, $\ldots$, $x_{n1},\ldots,x_{nt}$. Let S consist of the vectors $(1,x_{2i_2}, x_{3i_3},\ldots,x_{ni_n})$ in R^n with $1 \leqq i_2 \leqq t,\ldots,1 \leqq i_n \leqq t$. It can be shown that S is regular. We have $\dim S = n$, $z(S) = t^{n-1}$ and $m_{n-1} = t^{n-2}$, whence with $r = n - 1 = \dim S - 1$, $m_r = m_{n-1} = z(S)^{(n-2)/(n-1)} = z(S)^{(r-1)/r}$.

5. Proof of Theorem 3.

We have $N(x) = \mathfrak{N}(\alpha_1 x_1 + \ldots + \alpha_n x_n) = \ell^{(1)}(\underline{x})\ldots\ell^{(d)}(\underline{x})$ where $\ell^{(1)}(\underline{x}),\ldots,\ell^{(d)}(\underline{x})$ are the conjugates of the linear form $\ell(\underline{x}) = \alpha_1 x_1 + \ldots + \alpha_n x_n$.

We shall apply Theorem A to the d linear forms $\ell_1 = \ell^{(1)},\ldots,\ell_d = \ell^{(d)}$. We have to prove that under the hypothesis of Theorem 3, a constant η occuring in inequality (11) of Theorem A has $\eta \leqq d'$. Hence in view of Theorem A we have to show that

(18) $\qquad r < t$ implies $tm \leqq d'r$.

So let t and a rational subspace R^t of dimension t be fixed. The linear functionals on R^t form a vector space V. (It is best to make V into a real vector space if the field K is real, and a complex vector space otherwise). The restrictions $\hat{\ell}^{(1)},\ldots,\hat{\ell}^{(d)}$ of the given linear forms $\ell^{(1)},\ldots,\ell^{(d)}$ to R^t are elements of V. Let S consist of $\hat{\ell}^{(1)},\ldots,\hat{\ell}^{(d)}$. Since the given linear forms $\ell^{(1)},\ldots,\ell^{(d)}$ have rank n (which follows from the fact that the coefficients of $\ell(\underline{\underline{x}})$ can be extended to a basis of K), their restrictions to R^t have rank t, and we have $\dim S = t$. We have $d > n > 1$, whence $z(S) = d \geqq 3$.† Every element ϕ of the Galois group G_K of the least normal extension of the field K permutes $\ell^{(1)},\ldots,\ell^{(d)}$, and therefore permutes the elements of S. Moreover, if functionals $\hat{\ell}^{(v_1)},\ldots,\hat{\ell}^{(v_m)}$ are linearly dependent, then so are their images under ϕ . Thus ϕ induces an isomorphism of S in the sense defined in §4.

† This is true for $t \geqq 2$, because in this case $\hat{\ell}^1,\ldots,\hat{\ell}^d$ are distinct, since K is primitive. It need not be true for $t=1$, but then $r<t$ gives $r=0$, $m=0$, and so (18) holds. Assume $t \geqq 2$ in the sequel.

Now suppose $K = \underset{=}{Q}(\alpha)$, say. Let A be a subset of $\alpha^{(1)},\ldots,\alpha^{(\bar{d})}$ with $1 < z(A) < d$. We claim there is a ϕ in the Galois group G_K with

$$\phi(A) \neq A, \quad \phi(A) \cap A \neq \emptyset. \tag{19}$$

For otherwise, we have $d = pz(A)$ and a decomposition of $\{\alpha^{(1)},\ldots,\alpha^{(d)}\}$ into disjoint subsets $A = A_1, A_2, \ldots, A_p$, such that elements of G_K permute these p sets. The symmetric rational functions in the elements of A form a field H. It is clear that elements of H have at most p distinct conjugates, so that $\deg H \leqq p$. On the other hand, $\alpha = \alpha^{(1)}$ is algebraic over H of degree $\leqq z(A)$. Since α is algebraic over the rationals of degree $d = pz(A)$, H must be a subfield of K, of degree p. Since $1 < d/z(A) = p < d$, H is a non-trivial subfield; a contradiction. Hence there is in fact a ϕ with (19).

Since every ϕ in G_K induces an isomorphism of S, it follows that S is regular, and Lemma 3 applies. If $r = 0$, we have $m = m_0 = 0$, whence $tm = d'r = 0$. If $r > 0$, we have $m \leqq m_r \leqq 2^{(r-1)/2} d^{(r-1)/2}$,

$$m/r \leqq 2^{(r-1)/2}\, d^{(r-1)/r}/r = f(r) ,$$

say. Now since $d \geqq 3$, the function $f(r)$ is increasing as r runs through the positive integers, and since $r < t$, we obtain $m/r \leqq f(t-1)$. Since $t \leqq n$, we further obtain

$$tm/r \leq tf(t-1) \leq nf(n-1)$$

$$= \frac{n}{n-1} 2^{(n-2)/2} d^{(n-2)/(n-1)} = d' .$$

Thus (18) holds.

REFERENCES

[1] Z.I. Borevich and I.R. Shafarevich, Number theory, (Translated from the 1964 Russian ed.) Academic Press, New York and London, 1966.

[2] M. Fujiwara, On some applications of W.M. Schmidt's theorem, Michigan Math. J. (To appear).

[3] K.F. Roth, Rational approximations to algebraic numbers, Mathematika 2 (1955), pp. 1-20.

[4] W.M. Schmidt, Linearformen mit algebraischen Koeffizienten, II, Math. Ann. 191 (1971), pp. 1-20.

[5] W.M. Schmidt, Norm form equations, Annals of Math. (To appear).

[6] E. Wirsing, On approximations of algebraic numbers by algebraic numbers of bounded degree, Proc. of Symp. in Pure Math. XX (1971), (1969 Number Theory Institute) pp. 213-247.

University of Colorado, Boulder, Colorado

FURTHER ADVANCES IN THE THEORY OF LINEAR FORMS IN LOGARITHMS

H.M. Stark*

1. Introduction

Let $\alpha_1,\ldots,\alpha_n$ be non-zero algebraic numbers of degree less than or equal to d and heights (the maximum absolute value of the coefficients in the minimal defining polynomial) less than or equal to $A_1,\ldots,A_n$ respectively, where each $A_i \geq e$. Let $\beta_1,\ldots,\beta_n$ denote algebraic numbers of degrees less than or equal to d and heights less than B. In recent years, many difficult problems in number theory have been successfully solved by reducing them to a study of inequalities of the form

$$(1) \qquad 0 < |\beta_1 \log \alpha_1 + \ldots + \beta_n \log \alpha_n| < e^{-\delta H}$$

where, for convenience, we assume throughout that the logarithms are principal valued, $\delta > 0$ and $H > B$. In the applications an explicit upper bound for H in terms of $A = \max_{1\leq j\leq n} A_j$, d, n, and δ is desired. It was proved in [1,IV] that when the β_j are rational integers,

*This work was supported in part by NSF Grant No. P 28969.

$$H < c(\log A)^{(2n+1)^2},$$

where $c = c(n,d,\delta)$ was explicitly given.

Many of the applications involve cases where $A = A_n$ is much larger than $A_1,\ldots,A_{n-1}$. Much work has been done on this case recently. If we let $A' = \max\limits_{1\leqslant j\leqslant n-1} A_j$, then it was proved in [2] that if $\varepsilon > 0$

$$H < c(\log A)^{1+\varepsilon}$$

where now $c = c(n,d,A',\delta,\varepsilon)$ can be given effectively in terms of $n,d,A',\delta,\varepsilon$ and indeed, even if only $B < H^{\log H}$. Up to the problem of removing the ε, this is the best possible result.

The method of [2] develops considerably further the idea introduced in [4] in which an auxiliary function $\psi(z)$ is ultimately shown to vanish at $z = 1/p$ for a large prime p and then the desired goal is reached by means of algebraic lemmas. In the case that all the A_j are roughly the same size, the interpolation part of [2] can be seen to be workable only for $H<c(\log A)^{n+\varepsilon}$ and to fail if $H = c(\log A)^{\kappa}$, $\kappa < n$. In actually attempting to achieve anything better than $H < c(\log A^{n^2+\varepsilon}$ by the method of [2], we find that we cannot get p to be large enough, but even if we could, we would end up with a sequence of algebraic numbers of increasing heights rather than decreasing heights (the $\alpha_n^{(j)}$ of [2]). Thus the main innovations of this paper lie in Sections 3 and 5 where we develop the algebraic techniques considerably further. In particular, we prove

THEOREM 1. Suppose that each $A_j < A$, $B < H^{\log H}$, $\varepsilon > 0$, $\delta > 0$. If (1) holds then $H < c(\log A)^{n+\varepsilon}$ where $c = c(n,d,\varepsilon,\delta)$ is effectively computable.

Although Theorem 1 seems to be the best result attainable using the present interpolation techniques, I know of no example contradicting the conjecture that $H < c(\log A)^{1+\varepsilon}$. For this reason, the problem of removing the ε does not seem too important unless an application comes to light requiring it. Theorem 1 is actually a special case of the following result which also includes the result of [2] as a special case.

THEOREM 2. Suppose that $B < H^{\log H}$, $\varepsilon > 0$, $\delta > 0$. If (1) holds then

$$H < c\left(\prod_{j=1}^{n} \log A_j\right)^{1+\varepsilon}$$

where $c = c(n,d,\varepsilon,\delta)$ is effectively computable.

The method of proof of Theorem 2 allows us to consider the case that the linear form in (1) equals 0. Thus we will prove the following which contains Theorem 2.

THEOREM 3. Assume that not all β_j are 0 and suppose that $B < H^{\log H}$, $\varepsilon > 0$, $\delta > 0$ and

$$|\beta_1 \log \alpha_1 + \ldots + \beta_n \log \alpha_n| < e^{-\delta H}. \tag{2}$$

There is an effectively computable constant $c = c(n,d,\varepsilon,\delta)$ such that if $H \geq c\left(\prod_{j=1}^{n} \log A_j\right)^{1+\varepsilon}$

then

$$\beta_1 \log \alpha_1 + \ldots + \beta_n \log \alpha_n = 0$$

and there are integers $j_1, \ldots, j_n$, not all zero, such that

$$\alpha_1^{j_1} \ldots \alpha_n^{j_n} = 1$$

and such that $|j_i| < H^{2D-1}$ $(1 \leq i \leq n)$ where D is the degree of $K = Q(\alpha_1, \ldots, \alpha_n, \ \beta_1, \ldots, \beta_n)$.

The bound $H^{\log H}$ for B was chosen because it is sufficient for all applications of which I am aware. However, it will be clear that the same proof (and the same choice of variables) works for $B < \exp(H^{\kappa})$ where $\kappa < \varepsilon/(8n)$ and c depends on κ as well as $n, d, \varepsilon, \delta$.

2. Some Lemmas From Other Sources. In this section we collect those results that we will use from other sources. When we speak of the height of an algebraic number α, we shall always mean the maximum absolute value of the coefficients in the minimal defining polynomial for α. We will, however, have occasion in this and the next section to speak of the maximum absolute value of α and its conjugates which we will denote by $h(\alpha)$.

We begin with a fundamental result in diophantine approximation which we will use in the following form.

DIRICHLET'S THEOREM. Given m real numbers $\gamma_1, \ldots, \gamma_m$ and a real number $q \geq 1$, there exist integers $q_1, \ldots, q_{m+1}$ with $0 < q_{m+1} \leq q$ such that

$$|q_{m+1}\gamma_j - q_j| \leqslant q^{-1/m}, \qquad 1 \leqslant j \leqslant m.$$

LEMMA 1. Let M and N be integers, $N > M > 0$ and let u_{jk} be integers $(1 \leqslant j \leqslant M,\ 1 \leqslant k \leqslant N)$ with absolute values at most $U \geqslant 1$. Then there exist integers $x_1, \ldots, x_n$ not all zero with absolute values at most $(NU)^{M/(N-M)}$ such that

$$\sum_{k=1}^{N} u_{jk} x_k = 0, \qquad 1 \leqslant j \leqslant M .$$

This is Lemma 1 of [1,I].

LEMMA 2. Let α be a non-zero algebraic integer in a field K of degree D. There exists an effectively computable constant $c_D > 0$ (depending only on D) such that if $h(\alpha) \leqslant 1+c_D$ then α is a root of unity.

Proof. This is an improvement of a theorem of Kronecker. The best known value of c_D is $c_D = (30\ D^2 \log 6D)^{-1}$ which is given in [3]. That c_D exists however, is easily seen. Given $\delta > 0$, by Dirichlet's Theorem there exists an integer n, $0 < n \leqslant \delta^{-D}$, such that any conjugate of α^n has argument between $-2\pi\delta$ and $2\pi\delta$. With $c_D < 2^{1/D} - 1$, we see that the norm of α is ± 1 and hence $|\alpha^{(\sigma)}| \geqslant (1+c_D)^{1-D}$, where $\alpha^{(\sigma)}$ denotes any conjugate of α. Therefore

$$(1+c_D)^{-D\delta^{-D}} \leqslant |\alpha^{(\sigma)n}| \leqslant (1+c_D)^{\delta^{-D}}.$$

Hence if δ and then c_D are chosen sufficiently small, we must have

$$\prod_{(\sigma)} (x-\alpha^{(\sigma)n}) = \sum_{k=0}^{D} \binom{D}{k}(-1)^k x^{D-k} = (x-1)^D$$

and α is an nth root of unity.

LEMMA 3. If α is an algebraic number of height A and degree less than or equal to D and a is the leading coefficient of the minimal defining polynomial for α then given a rational integer $s \geqslant 0$, there exist rational integers $a_t(s)$ with absolute values less than or equal to $(2A)^s$ and such that

$$(a\alpha)^s = \sum_{t=0}^{D-1} a_t(s)\alpha^t$$

This is proved on p. 206 of [1,I].

LEMMA 4. Let K be an algebraic number field of degree D and α be an element of K of height A. Then $h(\alpha) \leqslant DA$ and if $\alpha \neq 0$, $h(\alpha^{-1}) \leqslant DA$.

This is proved on p. 206 of [1,I].

LEMMA 5. Let $\alpha_1,\ldots,\alpha_n$ be non-zero elements of an algebraic number field K and let $\alpha_1^{1/p},\ldots$ $\alpha_n^{1/p}$ denote fixed p^{th} roots for some prime p. Further let $K' = K(\alpha_1^{1/p},\ldots,\alpha_{n-1}^{1/p})$. Then either $K'(\alpha_n^{1/p})$ is an extension of K' of degree p or we have

$$\alpha_1^{j_1}\ldots\alpha_n^{j_n} = \eta^p$$

for some η in K and some integers $j_1, \ldots, j_n$ with $0 \leqslant j_\ell < p$ $(1 \leqslant \ell \leqslant n-1)$ and $j_n = 1$.

This is Lemma 3 of [2].

LEMMA 6. Let $f(z)$ be an entire function and

$$F(z) = \{(z-\gamma_1)\ldots(z-\gamma_R)\}^{S+1}$$

where R and S are integers, $R \geqslant 1$, $S \geqslant 0$, and $\gamma_1, \ldots, \gamma_R$ are distinct complex numbers. Suppose $\gamma \neq \gamma_r$ $(1 \leqslant r \leqslant R)$. Let Γ and Γ_r $(1 \leqslant r \leqslant R)$ be circles oriented in the positive direction such that Γ contains all the γ_r and γ in its interior while Γ_r contains γ_r but no other γ_k nor γ in its interior. Put $f_m(z) = (d^m/dz^m)f(z)$. Then

$$\frac{1}{2\pi i}\int_\Gamma \frac{f(z)}{(z-\gamma)F(z)}$$

$$= \frac{f(\gamma)}{F(\gamma)} + \frac{1}{2\pi i}\sum_{r=1}^{R}\sum_{m=0}^{S}\frac{f_m(\gamma_r)}{m!}\int_{\Gamma_r}\frac{(z-\gamma_r)^m}{(z-\gamma)F(z)}\,dz \, .$$

This is proved on p. 212 of [1,I].

3. Some Further Algebraic Lemmas. In this section we develop the necessary algebraic machinery for our results. The following lemma may be of independent interest.

LEMMA 7. Let K be an algebraic number field of degree D and let $\zeta_1, \ldots, \zeta_m$ be multiplicatively independent units of K such that $h(\zeta_i) \leqslant A, i = 1, \ldots, m$. Suppose that ζ_{m+1} is a unit of K with $h(\zeta_{m+1}) \leq A$ such that $\zeta_1,\ldots,\zeta_{m+1}$ are multiplicatively dependent. Then there is an effectively computable constant $c_1 = c_1(D) > 1$ depending only upon D with the following property: If $t_1,\ldots,t_{m+1}$ are relatively prime integers such that $\zeta_1^{t_1}\cdots\zeta_{m+1}^{t_{m+1}}$ is a root of unity (such integers exist) then

$$|t_i| \leq (c_1 \log A)^m \quad \text{for} \quad 1 \leq i \leq m+1 .$$

<u>Proof</u>. We will show that

$$c_1 = D^2/\log(1+c_D)$$

will do where c_D is the constant in Lemma 2. Not only do relatively prime integers $t_1,\ldots,t_{m+1}$ exist such that $\zeta_1^{t_1}\cdots\zeta_{m+1}^{t_{m+1}}$ is a root of unity, they are unique up to multiplication by -1 for otherwise there would be a dependence among $\zeta_1,\ldots,\zeta_m$. Suppose that

$$|t_I| = \max_{1\leq i\leq m+1} |t_i| > (c_1 \log A)^m$$

By Dirichlet's Theorem, there exist relatively prime integers $q_1,\ldots,q_{m+1}$ with $0<q_I\leq q = (c_1 \log A)^m$ such that

$$(3) \quad \left|q_I \frac{t_i}{t_I} - q_i\right| \leq q^{-1/m} \quad 1\leq i<I \quad \text{and} \quad I<i\leq m+1$$

and of course this holds for $i = I$ also (note $q > 1$ by Lemma 2). Let

$$\zeta = \zeta_1^{q_1}\cdots\zeta_{m+1}^{q_{m+1}}$$

Since the norm of each ζ_1 is ± 1 we see that $h(\zeta_i^{-1}) \leq A^{D-1}$. Thus $|\log|\zeta_i|| \leq (D-1)\log A$ (since $D \geq 2$ as otherwise the only units are ± 1). We see that by the choice of the t_i,

$$\begin{aligned} |\log|\zeta|| &= \left|\sum_{i=1}^{m+1}\left\{\left(q_i - q_I \frac{t_i}{t_I}\right) + \frac{q_I}{t_I} t_i\right\} \log|\zeta_i|\right| \\ &= \left|\sum_{i=1}^{m+1}\left(q_i - q_I \frac{t_i}{t_I}\right) \log|\zeta_i|\right| \\ &\leq \frac{(m+1)(D-1)\log A}{q^{1/m}} < \log(1+c_D) \end{aligned}$$

(note that $m+1 \leq D$). In particular, $|\zeta| < 1+c_D$ and the same is seen to be true of every conjugate of ζ. By Lemma 2, ζ is a root of unity. But $0 < q_I < |t_I|$ and this contradicts the uniqueness of the t_i and proves the lemma.

<u>LEMMA 8</u>. Let K be an algebraic number field of degree D and let $\eta_1, \ldots \eta_s$ be multiplicatively independent units of K such that $h(\eta_i) \leq A$, $i = 1, \ldots, s$. There exists a system of fundamental units $\varepsilon_1, \ldots, \varepsilon_r$ (so that $r \geq s$) of K and an effectively computable number $c_2 = c_2(D) > 1$ depending only upon D such that

$$\eta_i = \omega_i \prod_{j=1}^{i} \varepsilon_j^{e_{ij}}, \qquad 1 \leq i \leq s \tag{4}$$

with the ω_i being roots of unity in K and

$$|e_{ij}| \leq (c_2 \log A)^i \tag{5}$$

Proof. That a system of fundamental units satisfying (4) exists is well known and then we may easily alter it so as to have each $e_{ii} > 0$ and

$$-e_{ii} < e_{ij} \leqslant 0 , \qquad 1 \leqslant j < i .$$

Assume for $I \leqslant s$ that we have

$$(6) \qquad h(\varepsilon_i) \leqslant A^{2^{i-1}} , \qquad 1 \leqslant i \leqslant I-1$$

this being true for $I = 2$ since $h(\varepsilon_1) \leqslant h(\eta_1) \leqslant A$. Then

$$\varepsilon_I^{e_{II}} = \eta_I \prod_{j=1}^{I-1} \varepsilon_j^{-e_{Ij}}$$

and hence

$$h(\varepsilon_I) \leqslant A \prod_{j=1}^{I-1} A^{2^{j-1}} \leqslant A^{2^{I-1}}$$

Therefore (6) is true for $1 \leqslant i \leqslant s$ by induction and in particular since $s < D$,

$$h(\varepsilon_i) \leqslant A^{2^D} , \qquad 1 \leqslant i \leqslant s .$$

We now apply Lemma 7 to (4) with $\zeta_j = \varepsilon_j$, $1 \leqslant j \leqslant i$, $\zeta_{i+1} = \eta_i$, $m = i$, $t_{i+1} = 1$. The result is

$$|e_{ij}| \leqslant (c_1 2^D \log A)^i , \qquad 1 \leqslant j \leqslant i$$

and this proves the lemma with $c_2 = 2^D c_1$.

LEMMA 9. Let S be a set of primes with minimal element p_o. Let $\alpha_1,\ldots,\alpha_n$ be $n > 1$ non-zero algebraic numbers with heights less than $A_1,\ldots,A_n$ respectively where $A_i \geqslant e$ $(1\leqslant i\leqslant n)$ and suppose that $\alpha_1,\ldots,\alpha_n$ all belong to a field K of degree D. Suppose further that if p is in S then there are integers $j_1,\ldots,j_n$, $0 < j_i < p$ $(1 \leqslant i \leqslant n)$, $j_n \neq 0$ with

$$\alpha_1^{j_1}\ldots\alpha_n^{j_n} = \eta^p \tag{7}$$

where η is in K. Then with at most $(\log p_0)^{-1} \cdot \log[(2Dn)^n \prod_{i=1}^{n} \log A_i]$ exceptions p, if p is in S then there are integers $j_1,\ldots,j_n$,

$$|j_i| \leqslant p(2Dn)^n \prod_{\ell=1}^{n} \log A_\ell \qquad (1 \leqslant i \leqslant n) ,$$

$j_n \not\equiv 0 \pmod p$, such that an equation of the form (7) holds where η is a unit in K.

Proof. Suppose $\mathfrak{p}_1,\ldots,\mathfrak{p}_s$ are all of the prime ideals of K involved in the factorizations of the α_i. We will take $s > 0$ since otherwise the lemma is trivial. We may then write the principal ideals (α_i) as

$$(\alpha_i) = \prod_{\ell=1}^{s} \mathfrak{p}_\ell^{e_{i\ell}}$$

where $e_{i\ell}$ are rational integers. If $e_{i\ell} < 0$ and a_i is the leading coefficient in the defining polynomial for α_i then $a_i\alpha_i$ is an algebraic integer and hence $\mathfrak{p}_\ell^{-e_{i\ell}} | a_i$. Therefore, by taking the norm from K to Q,

$$|e_{i\ell}| \leqslant \frac{D}{\log 2} \log A_i \leqslant 2D \log A_i . \tag{8}$$

In like manner, we may look at α_i^{-1} and find that (8) is true if $e_{i\ell} > 0$.

Let the vector $e_i = (e_{i1},\ldots,e_{is})$, $i \leqslant i < n$, and suppose that exactly r of the e_i are linearly independent over Q (by our assumption on s, $1 \leqslant r \leqslant n$). We temporarily renumber the α_i and p_ℓ so that

$$\Delta = \det(e_{i\ell}) \neq 0 \qquad (1 \leqslant i \leqslant r\ ,\ \ 1 \leqslant \ell \leqslant r).$$

Let

$$e_i = f_{i1}e_1 + \ldots + f_{ir}e_r\ , \qquad r+1 \leqslant i \leqslant n\ .$$

Then for rational integers $k_1,\ldots,k_n$, $\prod_{i=1}^{n} \alpha_i^{k_i}$ is a unit if and only if

$$(9) \qquad (k_1,\ldots,k_r) = -(k_{r+1},\ldots,k_n)F$$

where F is the matrix $(f_{i\ell})$ $r+1 \leqslant i \leqslant n$, $1 \leqslant \ell \leqslant r$ (when $r=n$ we interpret (9) to mean all the $k_i=0$). We may also estimate the $f_{i\ell}$ and Δ since they are given by determinants,

$$|\Delta| < r! \prod_{i=1}^{r} (2D \log A_i)$$
$$\leqslant (2D)^n\ n! \prod_{i=1}^{n} \log A_i$$

and in like manner

$$|\Delta f_{i\ell}| < (2D)^n n! \prod_{i=1}^{n} \log A_i$$

where the $\Delta f_{i\ell}$ are integers.

Suppose p is a prime, $p \nmid \Delta$. Then we see that $\prod_{i=1}^{n} (\alpha_i)^{k_i}$ is the pth power of an ideal of K if and only if (9) holds (mod p). Let p be an element of S and suppose that $p \nmid \Delta$. With the integers $j_1,\ldots,j_n$ in the hypothesis, we have

$$\Pi\ \alpha_i^{\Delta j_i} = \eta^{\Delta p}$$

for some η in K. In particular, once we know such a p in S exists, we have $r < n$ since $\Delta j_i \not\equiv 0 \pmod p$ for some i. Let $k_i = \Delta j_i$ $(r+1 \leqslant i \leqslant n)$ and define $k_1,\ldots,k_r$ by (9). In particular the k_i are all integers and

$$k_i = \Delta j_i + ph_i$$

where the h_i are all integers.

Now,

$$\Pi\ \alpha_i^{k_i} = (\eta^{\Delta} \prod_{i=1}^{n} \alpha_i^{h_i})^p$$

is a unit by the definition of the k_i and hence is the pth power of a unit of K. Further,

$$|k_i| \leqslant p(2D)^n n! \prod_{i=1}^{n} \log A_i\ , \qquad r+1 \leqslant i \leqslant n,$$

and

$$|k_i| \leqslant np(2D)^n n! \prod_{i=1}^{n} \log A_i, \qquad 1 \leqslant i \leqslant r,$$

and hence

$$|k_i| \leqslant p(2Dn)^n \prod_{i=1}^{n} \log A_i\ , \qquad 1 \leqslant i \leqslant n.$$

We may now renumber the α_i as they were originally. We now have

$$k_n \equiv \Delta j_n \not\equiv 0 \pmod p.$$

Thus to complete the proof, we have to estimate the number of p in S such that $p|\Delta$. Suppose we have x elements of S dividing Δ. Then clearly

$$p_o^x \leqslant \Delta < (2Dn)^n \prod_{i=1}^{n} \log A_i$$

This is false if

$$x > \frac{\log\left[(2Dn)^n \prod_{i=1}^{n} \log A_i\right]}{\log p_o}$$

and completes the proof of the lemma.

<u>LEMMA 10</u>. Let S be a set of primes with minimal element p_o and maximal element P_o. Let $\alpha_1,\ldots,\alpha_n$ be $n \geqslant 1$ non-zero algebraic numbers with heights less than $A_1,\ldots,A_n$ respectively where $A_i \geqslant e(1\leqslant i\leqslant n)$ and suppose that $\alpha_1,\ldots\alpha_n$ all belong to a field K of degree D. Suppose further that if p is in S then there are integers $j_1,\ldots,j_n$, $0 \leqslant j_\ell < p$, $j_n\neq 0$, with

$$\alpha_1^{j_1}\ldots\ \alpha_n^{j_n} = \eta^p$$

where η is in K. There are two effectively determinable numbers $c_o = c_o(D,n)$ and $c = c(D,n)$ depending only upon D and n such that if S has more than

$$D + \frac{c + (D^2+1)\log(\prod_{i=1}^{n} \log A_i)}{\log p_o}$$

elements then there are integers $h_1,\dots,h_n$ with $h_n \neq 0$,

$$|h_\ell| \leq c_o P_o^D \left(\prod_{i=1}^{n} \log A_i \right)^{2D-1}, \qquad 1 \leq \ell \leq n ,$$

such that

$$\alpha_1^{h_1} \dots \alpha_n^{h_n} = 1 .$$

Proof. By Lemma 9, there is a subset of S' of S such that if p is in S' then there are integers $k_1,\dots,k_n$ with $k_n \not\equiv 0$, (mod p),

$$(10) \qquad |k_\ell| \leq p(2Dn)^n \prod_{i=1}^{n} \log A_i \qquad (1 \leq \ell \leq n) ,$$

and such that

$$(11) \qquad \alpha_1^{k_1} \dots \alpha_n^{k_n} = \eta^p$$

where η is a unit in K. Further S' contains all the elements of S with at most $(\log p_o)^{-1} \cdot \log[(2Dn)^n \prod_{i=1}^{n} \log A_i]$ exceptions. There are at most $D - 1$ multiplicatively independent units among the η's ; let $p_1,\dots,p_s$ be elements of S' such that the units $\eta_1,\dots,\eta_s$ corresponding to them are independent but if p is in S' then the unit η corresponding to it is dependent upon $\eta_1,\dots,\eta_s$. We let $k_{i1},\dots,k_{in}$ be the exponents in (11) corresponding to p_i and η_i.

The case $s = 0$ means all the η's are roots of unity and so we presume for the moment $s \geq 1$. For p in S' we have integers $k_1, \ldots, k_n$ and a unit η corresponding to it and by assumption, there is a relation

$$\eta_1^{t_1} \cdots \eta_s^{t_s} \eta^t = \omega \tag{12}$$

where $t_1, \ldots, t_s, t$ are relatively prime integers and ω is a root of unity. Let

$$A = \exp\left\{\left[(2Dn)^n \prod_{i=1}^{n} \log A_i\right] \sum_{i=1}^{n} \log(DA_i)\right\}$$

$$< \exp\left[(2Dn)^{n+1} \left(\prod_{i=1}^{n} \log A_i\right)^2\right].$$

We see from Lemma 4, (10) and (11) that

$$h(\eta) \leq A, \quad h(\eta_i) \leq A, \quad 1 \leq i \leq s.$$

By Lemma 7,

$$|t| \leq (c_1 \log A)^{D-1}, \quad |t_i| \leq (c_1 \log A)^{D-1},$$

$$1 \leq i \leq s.$$

By Lemma 8, there is a system of fundamental units $\varepsilon_1, \ldots, \varepsilon_r$ of K $(r \geq s)$ such that

$$\eta_i = \omega_i \prod_{j=1}^{i} \varepsilon_j^{e_{ij}}, \qquad 1 \leq i \leq s$$

where the ω_i are roots of unity in K and

$$|e_{ii}| \leq (c_2 \log A)^i, \qquad 1 \leq i \leq s. \tag{13}$$

The unit η must depend only on $\varepsilon_1,\dots,\varepsilon_s$ so that we may set

$$\eta = \omega' \prod_{j=1}^{s} \varepsilon_j^{e_j}$$

where ω' is a root of unity in K. In order that (12) holds for integers $t_1,\dots,t_s,t$ with ω a root of unity, it is necessary and sufficient that

$$\sum_{i=j}^{s} e_{ij} t_i = -te_j \; , \qquad 1 \leqslant j \leqslant s \; .$$

In particular, for $t_1,\dots,t_s,t$ having no common factor, it is necessary that

$$t \mid \prod_{i=1}^{s} e_{ii} \; .$$

Suppose there is a prime p in S' such that $p \nmid e_{ii}$ for $1 \leqslant i \leqslant s$. Then $p \nmid t$ either. For this p we have

$$\alpha_1^{h_1} \dots \alpha_n^{h_n} = \omega^{p_1 \dots p_s p} \tag{14}$$

is a root of unity where

$$h_\ell = p_1 \dots p_s p \left(\sum_{i=1}^{s} p_i^{-1} t_i k_{i\ell} + p^{-1} t k_\ell \right), \qquad 1 \leqslant \ell \leqslant n \; .$$

In particular

$$h_n \equiv p_1 \dots p_s t k_n \not\equiv 0 (\mathrm{mod}\ p) \; ,$$

and thus certainly $h_n \neq 0$. Further since $s < D$,

$$|h_\ell| \leqslant P_o^D . D(c_1 \log A)^{D-1}(2Dn)^n \prod_{i=1}^{n} \log A_i$$

(15) $$\leqslant P_o^D c_1^{D-1}(2Dn)^{2nD}\left(\prod_{i=1}^{n} \log A_i\right)^{2D-1}$$

Since a primitive m^{th} root of unity generates an extension of Q of degree $\phi(m) > \frac{1}{2}m^{1/2}$, we see that we may eliminate the ω in (14) by raising both sides of (14) to a power less than $4D^2$. In case $s = 0$, we get an estimate far better than (15) by taking $h_\ell = k_\ell$ for any p in S'.

Thus to conclude the proof of the lemma, we must show that in the case $s \geqslant 1$, there is a p in S' such that $p \nmid e_{ii}$, $1 \leqslant i \leqslant s$. By (13)

$$\left|\prod_{i=1}^{s} e_{ii}\right| \leqslant (c_2 \log A)^{s(s+1)/2} < (c_2 \log A)^{D^2/2}$$

and thus if S' contains more than $x + s$ elements where

$$x = \frac{D^2 \log[c_2(2Dn)^{n+1}(\prod_{i=1}^{n} \log A_i)^2]}{2 \log p_o}$$

$$> \frac{D^2 \log(c_2 \log A)}{2 \log p_o}$$

we will have such a prime p in S'. This will be automatic if S has more than

$$D + (2 \log p_o)^{-1}\left\{2 \log[(2Dn)^n \prod_{i=1}^{n} (\log A_i)] + \right.$$

$$\left. + D^2 \log[c_2 (2Dn)^{n+1} \prod_{i=1}^{n} (\log A_i)^2] \right\}$$

elements. The lemma follows and indeed with

$$c_o = 4D^2 c_1^{D-1} (2Dn)^{2Dn}$$

and

$$c = \tfrac{1}{2}[D^2(n+1)+2n]\log(2Dn) + \frac{D^2}{2} \log c_2 .$$

4. The Interpolation Argument.

The goal of this and the following section is to prove the following dependence result.

LEMMA 11. Let K be an algebraic number field of degree D containing the non-zero algebraic numbers $\alpha_1,\ldots,\alpha_n$ and the numbers $\beta_1,\ldots,\beta_{n-1}$. We suppose that the height of each α_i is less than A_i where $A_i \geq e$ and the height of each β_i is less than B. By $\log \alpha_i$, we mean the principal value and we assume that $|\log \alpha_i| \geq 1$ for each i. Suppose that for some $H > 0$, $\delta > 0$ and $\varepsilon, 0 < \varepsilon < 1$, we have

$$(16) \qquad |\beta_1 \log \alpha_1 + \ldots + \beta_{n-1} \log \alpha_{n-1} - \log \alpha_n| < e^{-\delta H}$$

with

$$(17) \qquad B < \exp[(\log H)^3]$$

and

$$\prod_{i=1}^{n} \log(DA_i) < H^{1-\varepsilon} \tag{18}$$

There exists an effectively computable bound $H_o = H_o(n,D,\varepsilon,\delta)$ depending only on $n, D\varepsilon, \delta$ such that if $H > H_o$ then there are integers $j_1, \ldots, j_n$ with $j_n \neq 0$, $|j_i| \leq \frac{1}{2}H^{2D-1} (1 \leq i \leq n)$, and

$$\alpha_n^{j_n} = \alpha_1^{j_1} \ldots \alpha_{n-1}^{j_{n-1}} .$$

We let $\varepsilon = 8n\zeta$ and set

$$h = H^{\zeta}, \qquad k = H^{1-4n\zeta}, \qquad w = kh^{-3}$$

$$L_j = [w/\log(DA_j)]$$

where [ ] denotes the greatest integer function. All inequalities will be valid for $H > H_o$ where H_o will be easily seen to be effectively computable at each step. Since each $A_j \geq e$, we see from (18) that

$$w \geq L_j > wH^{\varepsilon-1} - 1 = h^{4n-3} - 1 > \tfrac{1}{2}h^{4n-3}.$$

More important, we have

$$L_j \log(DA_j) \leq w, \qquad 1 \leq j \leq n .$$

We define an auxiliary function ψ as follows,

$$\psi(z; m_1, \ldots, m_{n-1}) = \sum_{\lambda=0}^{L} p(\lambda) \prod_{j=1}^{n} \alpha_j^{\lambda_j z} \prod_{j=1}^{n-1} \gamma_j^{m_j} .$$

Here

$$\sum_{\lambda=0}^{L} = \sum_{\lambda_1=0}^{L_1} \cdots \sum_{\lambda_n=0}^{L_n}$$

and we have put

$$\gamma_j = \lambda_j + \lambda_n \beta_j , \qquad 1 \leqslant j \leqslant n-1 .$$

The numbers $p(\lambda) = p(\lambda_1,\ldots,\lambda_n)$ are rational integers and we let

$$P = \max_{\lambda_1,\ldots,\lambda_n} |p(\lambda)|.$$

It will be assumed throughout that the m_j are non-negative rational integers. By α^z we mean the principal value, $e^{z\log \alpha}$. The aim of this section is the following interpolation result which we will use in the next section for more than one function ψ .

<u>LEMMA 12</u>. Suppose that the hypotheses of Lemma 11 are satisfied. Suppose further that $P < e^{hk}$ and that for some x_o, $0 < x_o \leqslant 1$, and c, $2^{-4n/\zeta} \leqslant c \leqslant 1$, we have

$$\psi(x_o+\ell;\ m_1,\ldots,m_{n-1}) = 0$$

for all integers ℓ, $0 \leqslant \ell \leqslant h-1$ and all $m_1,\ldots,m_{n-1}$ with $\sum_{j=1}^{n-1} m_j \leqslant ck$. Then for any rational number q/p, $0 \leqslant q/p \leqslant h$, with $p \leqslant h$ and integers $m_1,\ldots,m_{n-1}$ with $\sum_{j=1}^{n-1} m_j \leqslant 2^{-2n}ck$ we have

$$\psi(\tfrac{q}{p};\ m_1,\ldots,m_{n-1}) = 0 .$$

For the rest of this section, we will assume the hypotheses of Lemma 12 (and hence Lemma 11 also). For the proof of Lemma 12 we will need a related auxiliary function. Let

$$\phi_{m_1,\ldots,m_{n-1}}(z_1,\ldots,z_{n-1})$$
$$= \prod_{j=1}^{n-1} (\log \alpha_j)^{m_j} \sum_{\lambda=0}^{L} p(\lambda) \prod_{j=1}^{n-1} (\alpha_j^{\gamma_j z_j} \gamma_j^{m_j}).$$

We see that

$$\phi_{m_1,\ldots,m_{n-1}}(z_1,\ldots,z_{n-1})$$
$$= \frac{\partial^{m_1+\ldots m_{m-1}}}{\partial z_1^{m_1} \ldots \partial z_{n-1}^{m_{n-1}}} \phi_{0,\ldots,0}(z_1,\ldots,z_{n-1}) .$$

<u>LEMMA 13</u>. If $0 \leqslant x \leqslant h^{2n}$, $m_1+\ldots+m_{n-1} \leqslant k$, then

$$(19) \quad \begin{aligned} &|\phi_{m_1,\ldots,m_{n-1}}(x,\ldots,x) \\ &- \prod_{j=1}^{n-1} (\log \alpha_j)^{m_j} \psi(x;m_1,\ldots,m_{n-1})| < \exp(-\tfrac{1}{2}\delta H) , \end{aligned}$$

$$(20) \quad \begin{aligned} &|\prod_{j=1}^{n-1} (\log \alpha_j)^{-m_j} \phi_{m_1,\ldots m_{n-1}}(x,\ldots,x) \\ &- \psi(x;m_1,\ldots,m_{n-1})| < \exp(-\tfrac{1}{2}\delta H) . \end{aligned}$$

<u>Proof</u>. For $0 \leqslant y \leqslant e^{\delta H}$ we have

$$\begin{aligned} &|y(\beta_1 \log \alpha_1+\ldots+\beta_{n-1} \log \alpha_{n-1}) \\ &- y \log \alpha_n| \leqslant y e^{-\delta H} \leqslant 1 \end{aligned}$$

hence

$$\left|\prod_{j=1}^{n-1}\alpha_j^{\beta_j y}-\alpha_n^y\right| = \left|\int_{y\log\alpha_n}^{y\sum_{j=1}^{n-1}\beta_j\log\alpha_j} e^z dz\right|$$

$$\leqslant ye^{-\delta H}\cdot e^{|y\log\alpha_n|+1} .$$

Now $L_n x < e^{\delta H}$ and hence by Lemma 4,

$$\left|\prod_{j=1}^{n-1}\alpha_j^{\gamma_j x}-\prod_{j=1}^{n}\alpha_j^{\lambda_j x}\right|$$

$$= \left|\prod_{j=1}^{n-1}\alpha_j^{\lambda_j x}\right|\cdot\left|\prod_{j=1}^{n-1}\alpha_j^{\beta_j\lambda_n x}-\alpha_n^{\lambda_n x}\right|$$

$$< \prod_{j=1}^{n-1}(DA_j)^{L_j x}\cdot L_n xe^{-\delta H}\cdot e^{L_n x[\log(DA_n)+\pi]+1}$$

$$< \exp(-\tfrac{3}{4}\delta H) .$$

Hence

$$|\phi_{m_1,\ldots,m_{n-1}}(x,\ldots,x)$$

$$-\prod_{j=1}^{n-1}(\log\alpha_j)^{m_j}\psi(x;m_1,\ldots,m_{n-1})|$$

$$= \left|\prod_{j=1}^{n-1}(\log\alpha_j)^{m_j}\right|\cdot\left|\sum_{\lambda=0}^{L}p(\lambda)\left(\prod_{j=1}^{n-1}\alpha_j^{\gamma_j x}-\right.\right.$$

$$\left.\left.-\prod_{j=1}^{n}\alpha_j^{\lambda_j x}\right)\prod_{j=1}^{n-1}\gamma_j^{m_j}\right| \leqslant \prod_{j=1}^{n-1}[\log(DA_j)+\pi]^{m_j}\cdot$$

$$\cdot\prod_{j=1}^{n}(L_j+1)\cdot P\cdot\exp(-\tfrac{3}{4}\delta H)\cdot(2wBD)^k < \exp(-\tfrac{1}{2}\delta H),$$

which proves (19). Since each $|\log \alpha_j| \geq 1$, (20) is an immediate consequence.

LEMMA 14. If $m_1+\ldots+m_{n-1} \leq k$, then for any complex z,

$$|\phi_{m_1,\ldots,m_{n-1}}(z,\ldots,z)|$$
$$< \exp(2hk + 5nw|z|) .$$

Further if r and s are integers, $1 \leq s \leq h$, $0 \leq r/s \leq h^{2n}s^{-n}$, either $\psi(r/s;\, m_1,\ldots,m_{n-1}) = 0$ and

(21) $$|\phi_{m_1,\ldots,m_{n-1}}(\tfrac{r}{s},\ldots,\tfrac{r}{s})| < \exp(-\tfrac{1}{2}\,\delta H)$$

or $\psi(r/s;\, m_1,\ldots,m_{n-1}) \neq 0$ and

(22) $$|\phi_{m_1,\ldots,m_{n-1}}(\tfrac{r}{s},\ldots,\tfrac{r}{s})|$$
$$> \exp(-2Dhks^n - 2nDwrs^{n-1}) .$$

Proof. We note that

$$\sum_{j=1}^{n-1} \gamma_j z \log \alpha_j = \sum_{j=1}^{n-1} (\lambda_j + \lambda_n \beta_j) z \log \alpha_j$$
$$= z \sum_{j=1}^{n} \lambda_j \log \alpha_j + z\lambda_n \left(\sum_{j=1}^{n} \beta_j \log \alpha_j - \log \alpha_n \right) .$$

Therefore

$$\left|\sum_{j=1}^{n-1} \gamma_j z \log \alpha_j\right| \leq |z| n(1+\pi)w + |z| w e^{-\delta H}$$
$$\leq 5nw|z| .$$

The first part of the lemma is now clear. If $\psi(r/s;\ m_1,\ldots,m_{n-1}) = 0$ then (21) follows from Lemma 13.

So we now assume that $\psi = \psi(r/s;m_1,\ldots,m_{n-1}) \neq 0$. Let $a_j (1 \leq j \leq n)$ and $b_j (1 \leq j \leq n-1)$ be the leading coefficients in the minimal defining polynomials for α_j and β_j. Let $\ell = [r/s]+1 \leq r/s+1$. Then

$$\zeta = \prod_{j=1}^{n} a_j^{L_j \ell} \cdot \prod_{j=1}^{n-1} b_j^{m_j} \cdot \psi$$

is a non-zero algebraic integer lying in an extension field K' of K of relative degree $\leq s^n$ and hence the degree of ζ over Q is $\leq Ds^n$. If $\zeta^{(\sigma)}$ is any conjugate of ζ, we have

$$\begin{aligned}|\zeta^{(\sigma)}| &\leq \prod_{j=1}^{n} A_j^{L_j \ell} \cdot B^k \cdot \prod_{j=1}^{n}(L_j+1) \cdot P \cdot \\ &\quad \cdot \prod_{j=1}^{n}(DA_j)^{L_j r/s} \cdot (2wBD)^k \\ &\leq \exp(\tfrac{3}{2}\ hk + 2nwr/s).\end{aligned}$$

By looking at the norm of ζ, we see that

$$|\zeta| > [\exp(\tfrac{3}{2}\ hk + 2nwr/s)]^{-Ds^n+1}.$$

Since

$$\prod_{j=1}^{n} \alpha_j^{L_j \ell} \cdot \prod_{j=1}^{n-1} b_j^{m_j} < e^{nw(1+r/s)} B^k,$$

we see that

$$(23) \qquad |\psi| > \exp(-\tfrac{3}{2}\, Dhks^n - 2nDwrs^{n-1}) \ .$$

In particular

$$\Big|\prod_{j=1}^{n-1}(\log \alpha_j)^{m_j}\psi\Big| > 2\exp(-\tfrac{1}{2}\,\delta H)$$

and (22) follows from Lemma 13 and (23).

<u>LEMMA 15</u>. Let J be an integer $1 \leq J \leq 2n$ and suppose that for some x_1, $0 \leq x_1 \leq 1$ and c, $2^{-4n/\zeta} \leq c \leq 1$ we have

$$|\phi_{m_1,\ldots,m_{n-1}}(x_1+\ell,\ldots,x_1+\ell)| \leq \exp(-\tfrac{1}{2}\,\delta H)$$

for all integers ℓ, $0 \leq \ell \leq h^J - 1$, and $m_1,\ldots,m_{n-1}$ with $\sum_{j=1}^{n-1} m_j \leq 2^{1-J}ck$. Then for all z with $|z| \leq h^{J+1}$ and $m_1,\ldots,m_{n-1}$ with $\sum_{j=1}^{n-1} m_j \leq 2^{-J}ck$, we have

$$|\phi_{m_1,\ldots,m_{n-1}}(z,\ldots,z)| < \exp(-3Dh^J k)$$

<u>Proof</u>. We let

$$R = [h^J] \ , \qquad\qquad S = [2^{-J}ck] \ .$$

We assume $m_1,\ldots,m_{n-1}$ are fixed with $\sum_{j=1}^{n-1} m_j \leq S$ and let $f(z) = \phi_{m_1,\ldots,m_{n-1}}(z,\ldots,z)$, $f_m(z) = \frac{d^m}{dz^m} f(z)$. If $0 \leq m \leq S$ then

$$m + m_1 + \ldots + m_{n-1} \leq 2S \leq 2^{1-J}ck$$

and hence for $0 \leq \ell \leq h^{J-1}$ and $0 \leq m \leq S$,

$$|f_m(x_1+\ell)| \leqslant \Bigg| \sum_{\substack{g_1=0 \\ g_1+\dots+g_{n-1}=m}}^{m} \cdots \sum_{g_{n-1}=0}^{m} \frac{m!}{g_1!\dots g_{n-1}!} \cdot$$

$$(24)\cdot\ \phi_{m_1+g_1,\dots,m_{n-1}+g_{n-1}}(x_1+\ell,\dots,x_1+\ell)\Bigg|$$

$$\leqslant (n-1)^m e^{-\frac{1}{2}\delta H} < n^k\ e^{-\frac{1}{2}\delta H}.$$

Let

$$F(z) = \prod_{\ell=0}^{R-1} (z-x_1-\ell)^{S+1}.$$

We apply Lemma 6, with $\gamma = \zeta$, $|\zeta| = h^{J+1}$, Γ the circle centered at 0 with radius h^{J+2}, and Γ_ℓ centered at $\gamma_\ell = x_1+\ell$ with radius 1/2. Then

$$\frac{1}{2\pi i}\int_\Gamma \frac{f(z)}{(z-\zeta)F(z)}dz = \frac{f(\zeta)}{F(\zeta)} + \frac{1}{2\pi i}\sum_{\ell=0}^{R-1}\sum_{m=0}^{S}\frac{f_m(x_1+\ell)}{m!}\cdot$$

$$\cdot \int_{\Gamma_\ell} \frac{(z-x_1-\ell)^m}{(z-\zeta)F(z)}\ dz.$$

Hence by (24) and Lemma 14,

$$\left|\frac{f(\zeta)}{F(\zeta)}\right| \leqslant \left|\frac{1}{2\pi i}\int_\Gamma \frac{f(z)}{(z-\zeta)F(z)}\ dz\right| + R(S+1)n^k\ e^{-\frac{1}{2}\delta H}.$$

$$\cdot\ 8^{S+1} < 2\ \frac{\exp(2hk+5nwh^{J+2})}{(\frac{1}{2}h^{J+2})^{R(S+1)}} + \exp(-\tfrac{1}{4}\ \delta H)\ .$$

Therefore

$$|f(\zeta)| \leqslant 2\,\frac{\exp(2hk+5nwh^{J+2})}{(\frac{1}{4}h)^{R(S+1)}} + (2h^{J+1})^{R(S+1)} \cdot$$

$$\cdot\exp(-\tfrac{1}{4}\,\delta H) \leqslant 2\,\exp[2hk + 5nwh^{J+2} - \tfrac{1}{2}h^{J}\cdot$$

$$\cdot\; 2^{-J}ck\,\log(\tfrac{1}{4}h)] + \exp(-\tfrac{1}{8}\,\delta H) < \exp(-3Dh^{J}k)\;.$$

This proves the lemma for $z = \zeta$ and the result follows in general from the maximum modulus theorem.

<u>Proof of Lemma 12.</u> Suppose that for some x_1 and integer J, $0 < x_1 \leqslant 1$, $1 \leqslant J < 2n$ that

$$|\phi_{m_1,\ldots,m_{n-1}}(x_1+\ell,\ldots,x_1+\ell)| < \exp(-\tfrac{1}{2}\,\delta H)$$

for $0 \leqslant \ell \leqslant h^{J}-1$ and $m_1+\ldots+m_{n-1} \leqslant 2^{1-J}ck$. Note that this is true for $J = 1$ and $x_1 = x_o$ by hypothesis. By Lemma 15, we have

$$|\phi_{n_1,\ldots,m_{n-1}}(1+\ell,\ldots,1+\ell)| \leqslant \exp(-3Dh^{J}k)$$

$$< \exp(-2Dhk-2nDwh^{J+1})$$

for $0 \leqslant \ell \leqslant h^{J+1} - 1$ and $m_1+\ldots+m_{n-1} \leqslant 2^{-J}ck$. Thus by Lemma 14 with $r = 1+\ell$, $s = 1$, we have

$$|\phi_{m_1,\ldots,m_{n-1}}(1+\ell,\ldots,1+\ell)| < \exp(-\tfrac{1}{2}\,\delta H)$$

in this range. This is the same result with $J + 1$ in place of J and $x_1 = 1$ in place of x_1. Thus by induction and Lemma 15 with $J = 2n$,

$$|\phi_{m_1,\dots,m_{n-1}}(\tfrac{q}{p},\dots,\tfrac{q}{p})| < \exp(-3Dh^{2n}k)$$

$$< \exp(-2Dhkp^n - 2nDwqp^{n-1}) ,$$

for all q/p, $0 < q/p \leq h$, $1 \leq p \leq h$, and all $m_1,\dots,m_{n-1}$ with $m_1+\dots+m_{n-1} \leq 2^{-2n}ck$. Hence by Lemma 14 with $r = q$, $s = p$, we have

$$\psi(\tfrac{q}{p};\ m_1,\dots,m_{n-1}) = 0$$

in this range. This completes the proof of Lemma 12.

5. The Algebraic Reductions.

We continue to assume the hypotheses of Lemma 11 as well as the notation introduced before Lemma 12. We begin by constructing a particular function ψ.

LEMMA 16. There exists integers $p(\lambda)$, not all zero, with $P < e^{hk}$ such that for $1 \leq \ell \leq h$ and $\sum_{j=1}^{n-1} m_j \leq k$ we have

$$\psi(\ell;\ m_1,\dots,m_{n-1}) = 0 .$$

Proof. Let $a_i (1 \leq i \leq n)$ and $b_j (1 \leq j \leq n-1)$ be the leading coefficients of the minimal defining polynomials of α_i and β_j. By Lemma 3 for any $s \geq 0$ there are rational integers $a_{if}(s)$ and $b_{jg}(s)$ such that

$$(a_i\alpha_i)^s = \sum_{f=0}^{D-1} a_{if}(s)\alpha_i^f , \quad |a_{if}(s)| \leqslant (2A_i)^s ,$$

and

$$(b_j\beta_j)^s = \sum_{g=0}^{D-1} b_{jg}(s)\beta_j^g , \quad |b_{jg}(s)| \leqslant (2B)^s .$$

We have for $\ell \geqslant 0$

$$\prod_{i=1}^{n} a_i^{\ell L_i} \cdot \prod_{j=1}^{n-1} b_j^{m_j} \psi(\ell;\, m_1,\ldots,m_{n-1})$$

$$= \sum_{\lambda=0}^{L} p(\lambda) \prod_{i=1}^{n} a_i^{\ell(L_i-\lambda_i)} \cdot \prod_{i=1}^{n} \Big(\sum_{f_i=0}^{D-1} a_{if_i}(\lambda_i\ell)\alpha_i^{f_i} \Big) \cdot \tag{25}$$

$$\cdot \prod_{j=1}^{n-1} \left[\sum_{r_j=0}^{m_j} \binom{m_j}{r_j} (\lambda_j b_j)^{m_j-r_j} \lambda_n^{r_j} \sum_{g_j=0}^{D-1} b_{jg_j}(r_j)\beta_j^{g_j} \right].$$

In order that $\psi = 0$, it is sufficient that the coefficient of $\prod_{i=1}^{n} \alpha_i^{f_i} \cdot \prod_{j=1}^{n-1} \beta_j^{g_j}$ on the right side of (25) is zero for each of the D^{2n-1} possible choices of the f_i and g_j. This gives us the equations

$$\sum_{\lambda=0}^{L} p(\lambda) \prod_{i=1}^{n} a_i^{\ell(L_i-\lambda_i)} \cdot \prod_{i=1}^{n} a_{if_i}(\lambda_i\ell) \cdot \tag{26}$$

$$\cdot \prod_{j=1}^{n-1} \Big(\sum_{r_j=0}^{m_j} \binom{m_j}{r_j} (\lambda_j b_j)^{m_j-r_j} \lambda_n^{r_j} \, b_{jg_j}(r_j) \Big) = 0.$$

Even with $0 < \ell \leqslant h$, $0 \leqslant m_j \leqslant k$ $(1 \leqslant j \leqslant n-1)$, we have

$$M \leqslant D^{2n-1} h(k+1)^{n-1}$$

equations for the

$$N = \prod_{j=1}^{n} (L_j+1) > \frac{w^n}{\prod_{j=1}^{n} \log(DA_j)} > \frac{w^n}{H^{1-\varepsilon}}$$

$$\geq k^{n-1} h^n > 2M$$

unknowns $p(\lambda)$. Further, the coefficient of any $p(\lambda)$ in (26) is an integer and is bounded above in absolute value by

$$\prod_{i=1}^{n} A_i^{hL_i} \cdot \prod_{i=1}^{n} (2A_i)^{L_i h} \cdot \prod_{j=1}^{n-1} [2^{m_j}(L_j B)^{m_j} L_n^{m_j}(2B)^{m_j}]$$

$$< \exp(\tfrac{1}{2} hk) = U .$$

Therefore, by Lemma 1, there are integers $p(\lambda)$, not all zero, with $P < NU < e^{hk}$ such that (26) holds for $1 \leq \ell \leq h$, $0 \leq m_j \leq k$ and all choices of the f_i and g_j. This completes the proof of the lemma.

We take a particular choice of integers $p_o(\lambda)$, not all zero, satisfying Lemma 16 and let ψ_o be the corresponding function ψ. In particular we shall assume that $\Lambda_1,\ldots,\Lambda_n$ are chosen so that $p_o(\Lambda_1,\ldots,\Lambda_n) \neq 0$.

<u>LEMMA 17</u>. With the exception of at most $2/\zeta$ primes p, if p is a prime, $h^{1/2} \leq p \leq h$, then there are integers $j_1,\ldots,j_n$, $0 \leq j_i < p$ $(1\leq i\leq n)$, $j_n = 1$, such that

$$\alpha_1^{j_1}\ldots\alpha_n^{j_n} = \eta^p$$

where η is in K.

Proof. Indeed the $2/\zeta$ is slightly high. Suppose p is an exceptional prime, that is $h^{1/2} \leq p \leq h$ but there are not integers $j_1, \ldots, j_{n-1}$, $j_n = 1$ such that $\Pi\alpha_i^{j_i}$ is a pth power in K. Then by Lemma 5, $K'(\alpha_n^{1/p})$ is an extension of $K' = K(\alpha_1^{1/p}, \ldots, \alpha_{n-1}^{1/p})$ of degree p. For any function ψ satisfying the hypotheses of Lemma 12, we have

$$\psi(\tfrac{q}{p}, m_1, \ldots, m_{n-1}) = 0 \tag{27}$$

for any q with $0 \leq q/p \leq h$ and $m_1, \ldots m_{n-1}$ with $\sum_{j=1}^{n-1} m_j \leq 2^{-2n}ck$. Let us examine (27) more closely. Written out, it says that

$$\sum_{\lambda_n=0}^{L_n} [\sum_{\lambda_1=0}^{L_1} \cdots \sum_{\lambda_{n-1}=0}^{L_{n-1}} p(\lambda) \prod_{i=1}^{n-1} (\alpha_i^{\lambda_i q/p} \gamma_i^{m_i})](\alpha_n^{1/p})^{\lambda_n q}$$

$$= 0 .$$

If $q \not\equiv 0 \pmod{p}$ then we have the above sum still zero when λ_n is summed over any single congruence class (mod p). We take in particular the class $\lambda_n \equiv \Lambda_n \pmod{p}$ and get (for $p \nmid q$),

$$\sum_{\substack{\lambda_n=0 \\ \lambda_n \equiv \Lambda_n \pmod{p}}}^{L_n} [\sum_{\lambda_1=0}^{L_1} \cdots \sum_{\lambda_{n-1}=0}^{L_{n-1}} p(\lambda) \prod_{i=1}^{n-1} (\alpha_i^{\lambda_i q/p} \gamma_i^{m_i})](\alpha_n^{1/p})^{\lambda_n q} \tag{28}$$

$$= 0 .$$

This holds in particular for $q = 1+\ell p$ $(0 \leqslant \ell \leqslant h-1)$ and $\Sigma\, m_j \leqslant 2^{-2n} ck$.

Now suppose $p = p_1, \ldots, p_r$ are all exceptions to the lemma where $r = [2/\zeta]$. Given ψ_i with coefficients $p_i(\lambda)$ for $0 \leqslant i \leqslant r-1$, define ψ_{i+1} with coefficients $p_{i+1}(\lambda)$ by

$$p_{i+1}(\lambda_1, \ldots, \lambda_n) = \begin{cases} p_i(\lambda_1, \ldots, \lambda_n) & \text{if } \lambda_n \equiv \Lambda_n (\operatorname{mod} p_{i+1}) \\ 0 & \text{if } \lambda_n \not\equiv \Lambda_n (\operatorname{mod} p_{i+1}). \end{cases}$$

It follows from Lemma 12 and (28) by induction starting with ψ_o that for $1 \leqslant i \leqslant r$

$$\psi_i\left(\frac{1+\ell p_i}{p_i};\ m_1, \ldots, m_{n-1}\right) = 0 \ , \tag{29}$$

$$0 \leqslant \ell \leqslant h-1 \ , \qquad \Sigma\, m_j \leqslant 2^{-2ni} k \ .$$

We see that

$$p_r(\lambda) = \begin{cases} p_o(\lambda) & \text{if } \lambda_n \equiv \Lambda_n (\operatorname{mod} p_i), \text{ all } i,\ 1 \leqslant i \leqslant r \\ 0 & \text{otherwise} \end{cases}$$

But

$$p_1 \cdots p_r > h^{r/2} > (H^{\zeta/2})^{\frac{2}{\zeta}-1} > Hh^{-1} > L_n$$

and hence

$$p_r(\lambda) = \begin{cases} p_o(\lambda) & \text{if } \lambda_n = \Lambda_n \\ 0 & \text{otherwise} \ . \end{cases}$$

Thus with $\ell = 0$ and $i = r$ in (29), we have

$$(30) \qquad \sum_{\lambda_1=0}^{L_1} \cdots \sum_{\lambda_{n-1}=0}^{L_{n-1}} p_o(\lambda_1,\ldots,\lambda_{n-1},\Lambda_n)\cdot$$

$$\cdot\, \alpha_1^{\lambda_1/p} \cdots \alpha_{n-1}^{\lambda_{n-1}/p} \gamma_1^{m_1} \cdots \gamma_{n-1}^{m_{n-1}} = 0$$

and this is true for all $m_1,\ldots,m_{n-1}$ with $\Sigma\, m_j \leqslant 2^{-2nr}k$.

Thus, as in [2], we may peel off one summation sign at a time from (30) using the Vandermonde determinants,

$$\det(\gamma_j^{m_j}) = \det[(\lambda_j+\Lambda_n\beta_j)^{m_j}] \;,$$

$$0 \leqslant \lambda_j \leqslant L_j \;, \qquad\qquad 0 \leqslant m_j \leqslant L_j$$

until we finally arrive at $p_o(\lambda_1,\ldots,\lambda_{n-1},\Lambda_n) = 0$ for all $\lambda_1,\ldots,\lambda_{n-1}$. This contradicts the fact that $p_o(\Lambda_1,\ldots,\Lambda_n) \neq 0$ and proves the lemma.

<u>Proof of Lemma 11</u>. Let S be the set of primes p, $h^{1/2} \leqslant p \leqslant h$, for which there are integers $j_1,\ldots,j_n$, $0 \leqslant j_i \leqslant p(1\leqslant i\leqslant n)$, $j_n = 1$ such that $\prod_{i=1}^{n} \alpha_i^{j_i}$ is a pth power in K. By Bertrand's postulate and Lemma 17, S has at least

$$\left[\frac{\log(h^{1/2})}{\log 2}\right] - \frac{2}{\zeta} > \frac{\zeta}{4}\log H$$

elements. But Lemma 10 already applies to this situation when S has more than

$$D + \frac{c + (D^2+1)\log(H^{1-\varepsilon})}{\log(h^{1/2})} < \frac{2(D^2+2)}{\zeta}$$

elements. Therefore, by Lemma 10 there are integers $j_1,\ldots j_n$ with $j_n \neq 0$ such that

$$\alpha_1^{j_1}\ldots\alpha_n^{j_n} = 1$$

and for $1 \leqslant i \leqslant n$,

$$|j_i| < c_0 h^D (H^{1-\varepsilon})^{2D-1} < \tfrac{1}{2} H^{2D-1} .$$

This completes the proof of Lemma 11.

6. <u>Proof of Theorem 3</u>.

We assume that the hypotheses of Theorem 3 are satisfied by the numbers $\alpha_1,\ldots,\alpha_n$, $\beta_1,\ldots,\beta_n$ and that

$$H > c\Big(\prod_{j=1}^{n} \log A_j\Big)^{1+\varepsilon} \tag{31}$$

We may assume that the numbering is such that $\beta_n \neq 0$. Further, we may assume that $\alpha_1 = -1$, $A_1 = e$, $\beta_1 = 0$ since by adding these numbers to the linear form, none of our hypotheses are disturbed and D remains unchanged. However, we must now produce a relation involving only $\alpha_2,\ldots,\alpha_n$. If c is large then $H > c$ is also large. The argument that follows is valid for sufficiently large c where "sufficiently large" is effectively computable.

Let ε', $0 < \varepsilon' < 1$, be given by $1-2\varepsilon' = (1+\varepsilon)^{-1}$. We see from (31) that

$$\prod_{j=1}^{n} \log A_j < H^{1-2\varepsilon'}$$

and hence

$$\prod_{j=1}^{n} \log(DA_j) < H^{1-\varepsilon'} \tag{32}$$

We divide our linear form by β_n. By Lemma 4,

$$|\beta_n^{-1}| \leqslant BD < \exp(\tfrac{1}{2}\,\delta H)$$

and hence

$$\left|\frac{\beta_1}{\beta_n} \log \alpha_1 + \ldots + \frac{\beta_{n-1}}{\beta_n} \log \alpha_{n-1} + \log \alpha_n\right| < \exp(\tfrac{1}{2}\,\delta H) \tag{33}$$

By Lemma 4,

$$h\left(\frac{\beta_i}{\beta_n}\right) \leqslant (DB)^2 , \qquad 1 \leqslant i \leqslant n-1 ,$$

and there is an integer $b_i \neq 0$ such that $b_i\beta_i/\beta_n$ is an algebraic integer with $|b_i| \leqslant B^2$. If any α_i has $|\log \alpha_i| \leqslant 1$ then $|\log(-\alpha_i)| > 1$ and $-\alpha_i$ has the same height as α_i. Thus by changing the sign of some of the α_i we may guarantee that $|\log \alpha_i| \geqslant 1$ for all i while (32) remains unchanged and our linear form (33) becomes

$$|\beta_1' \log \alpha_1 + \ldots + \beta_{n-1}' \log \alpha_{n-1} - \log \alpha_n| < \exp(\tfrac{1}{2}\,\delta H) \tag{34}$$

where

$$\beta_i' = -\frac{\beta_i}{\beta_n}, \qquad 2 \leqslant i \leqslant n-1$$

and

$$\beta_1' = -\frac{\beta_1}{\beta_n} + \sum_{i=2}^{n-1} \varepsilon_i \frac{\beta_i}{\beta_n}$$

for some choice of $\varepsilon_1,\ldots,\varepsilon_n$ from the three values $-1,0,1$.

Therefore $h(\beta_i') < nD^2B^2$ $(1\leqslant i\leqslant n-1)$ and there is an integer $b_i' \neq 0$ such that $b_i'\beta_i'$ is an algebraic integer and $|b_i| \leqslant B^{2(n-1)}$. Thus each β_i' satisfies an equation of the form $\sum_{j=0}^{D} b_{ij}x^j = 0$ with

$$|b_{ij}| \leqslant B^{2(n-1)j}\binom{D}{j}(nD^2B^2)^j \leqslant (2nD^2)^D B^{2nD}.$$

In other words each β_i' has height less than $B' = (2nD^2)^D B^{2nD}$ which is certainly less than $\exp(\log H)^3)$. But now the hypotheses of Lemma 11 are satisfied and since $\alpha_1 = -1$, we see from Lemma 11 that there are integers $j_2,\ldots,j_n$ with $j_n \neq 0$, $|j_i| \leqslant H^{2D-1}(2\leqslant i\leqslant n)$ such that

$$\alpha_n^{j_n} = \alpha_2^{j_2}\ldots\alpha_{n-1}^{j_{n-1}}. \tag{35}$$

This is the desired multiplicative relation.

It remains to show that the linear form in (34) (and hence our original linear form) equals 0. If we take the logarithm of both sides of (35) we get

$$j_n \log \alpha_n = j_1 \log \alpha_1 + \ldots + j_{n-1} \log \alpha_{n-1}$$

where j_1 is an integer that compensates for using principal values and we have

$$|j_1| \leqslant \sum_{i=2}^{n} |j_i| < nH^{2D-1}.$$

Our linear form (34) may now be written as,

$$(36)\qquad \begin{aligned} &|\beta_1'' \log \alpha_1 + \beta_2'' \log \alpha_2 + \ldots + \beta_{n-1}'' \log \alpha_{n-1}| \\ &\qquad < e^{-\frac{1}{2}\delta H} \end{aligned}$$

where

$$\beta_i'' = \beta_i' - \frac{j_i}{j_n}, \qquad 1 \leqslant i \leqslant n-1 .$$

We see from the beginning of the last paragraph that

$$h(\beta_i'') < nD^2B^2 + H^{2D} < nD^2(B+H^D)^2$$

and there is an integer b_i'' such that $b_i''\beta_i''$ is an algebraic integer and $|b_i''| < H^{2D}B^{2(n-1)} < (B+H^D)^{2(n-1)}$. In other words each β_i'' has a height less than B'' where we may take

$$B'' < (2nD^2)^D(B+H^D)^{2nD} < \exp(3nD(\log H)^2) .$$

If some $\beta_i'' \neq 0$, $2 \leqslant i \leqslant n-1$, we will renumber the α_i so that $\beta_{n-1}'' \neq 0$ (and we still have $\alpha_1 = -1$). We divide (36) through by β_{n-1}'' and apply Lemma 11 again. We continue this reduction process, eliminating at least one logarithm at a time, until we arrive at a form

$$(37) \qquad |\beta \log \alpha_1| < \exp(-\frac{1}{2^{n-1}}\delta H)$$

with the height of β less than $\exp[(3nD)^{n-1}(\log H)^2]$. Further $\beta \log \alpha_1 = 0$ if and only if our original linear form is 0. But if $\beta \neq 0$, Lemma 4 contradicts the estimate (37) and hence $\beta = 0$. This completes the proof of Theorem 3.

Added in proof.

If nonzero algebraic numbers $\alpha_1,\ldots,\alpha_n$ are multiplicatively dependent, then we may use Lemma 10 to produce a dependence relation in which we can give an estimate of the exponents. For example, an application of Lemma 10 to the result of Theorem 3 allows us to replace D by the degree of $Q(\alpha_1,\ldots,\alpha_n)$ in the estimate of the j_i.

REFERENCES

[1] A. Baker, Linear forms in the logarithms of algebraic numbers, I, II, III, IV, Mathematika 13 (1966), 204-216, 14 (1967), 102-107, 220-228, 15 (1968), 204-216.

[2] A. Baker and H.M. Stark, On a fundamental inequality in number theory, Ann. Math. 94 (1971), 190-199.

[3] P.E. Blanksby and H.L. Montgomery, Algebraic integers near the unit circle, Acta Arithmetica 18 (1971), 355-369.

[4] H.M. Stark, A transcendence theorem for class-number problems, Ann. Math. 94 (1971), 153-173.

Massachusetts Institute of Technology, Cambridge, Massachusetts.

DIFFERENTIAL RINGS OF ANALYTIC FUNCTIONS OF A NONARCHIMEDEAN VARIABLE

E. G. Straus

1. Introduction.

In an earlier paper [1] we characterized the differential rings, R, of meromorphic functions of one complex variable which have the property that they are finite dimensional over the ring, L, of linear differential operators with constant coefficients; that is that the maximal number of elements of R which are linearly independent over L is bounded. It turned out that, if R is a ring of entire functions, then its elements are exponential polynomials, that is solutions of linear differential equations with constant coefficients. If R is a ring of meromorphic functions not all of which are entire then R is either a ring of rational functions or a ring of (singly or doubly) periodic functions.

In this note we prove the same result for functions of a nonArchimedean variable. Thus the domain, $\mathcal{D}$, of our functions is an algebraically

The preparation of this paper was supported in part by National Science Foundation Grant GP28696.

closed field of characteristic 0 with a non-Archimedean valuation and complete with respect to that valuation. Entire functions are represented by power series which converge for all values of the variable and meromorphic functions are ratios of entire functions. Since the rational integers form a bounded set it follows that the polynomials are the only entire exponential polynomials and that there are no nonconstant periodic meromorphic functions. Thus the analog to the results in [1] is

THEOREM 1.1 If $\mathcal{R}$ is a ring of functions meromorphic on $\mathcal{D}$ which is finite dimensional over $\mathcal{L}$, then $\mathcal{R}$ is a ring of rational functions and the set of all poles of elements of $\mathcal{R}$ is finite.

We prove this theorem in §2.

In §3 we show that the main lemma used in the proof of Theorem 1.1 does not generalize to linear operators with nonconstant coefficients. Thus additional tools would be needed to prove Theorem 1.1 when the operators in $\mathcal{L}$ are linear differential operators with polynomial coefficients.

2. Proof of Theorem 1.1

We need a few very obvious extensions from the theory of entire functions of a complex variable. The results for a nonArchimedean variable are simpler and more complete.

DEFINITION 2.1 Let $f(x) = \sum_{n=0}^{\infty} c_n x^n$ be entire on $\mathcal{D}$ then for $r \geq 0$ we have

$$M(r,f) = \sup_{|x|\leq r} |f(x)|, \quad m(r,f) = \max_n |c_n| r^n .$$

The function $\mu(\rho)$ where $\mu = \log M(r,f)$ and $\rho = \log r$ is the <u>maximum modulus diagram of</u> f.

We state the following well known facts without proof.

<u>LEMMA 2.2</u> For all $r > 0$ we have $M(r,f) = m(r,f)$. Thus the maximum modulus diagram of f is a convex polygonal curve whose edges have non-negative integral slopes, $\frac{d\mu}{d\rho} = \deg m(r,f)$.

<u>LEMMA 2.3</u> For all $r > 0$ and all entire f we have $M(r,f') \leq M(r,f)/r$.

<u>Proof</u>. By Lemma 2.2 we have

$$M(r,f') = m(r,f') = \max_n |n||c_n| r^{n-1}$$

$$\leq \max_n |c_n| r^{n-1} = m(r,f)/r = M(r,f)/r .$$

<u>MAIN LEMMA 2.4</u> Given a transcendental entire function f on $\mathcal{D}$, a nontrivial linear differential operator $L = c_k D^k + c_{k+1} D^{k+1} + \ldots + c_m D^m$, $c_k \neq 0$ and a monotonically increasing function ϕ so that

$$M(r,Lf) \leq \phi(M(r,f)) \quad \text{for all} \quad r \geq 0 . \tag{2.5}$$

Then

$$\int^{\infty} \frac{\phi(M)^{1/k} dM}{M^{1+1/k-\delta}} = \infty \quad \text{for all} \quad \delta > 0 . \tag{2.6}$$

Proof. According to Lemma 2.3 we have

$$\sup_{|x|\leq r} |c_{k+1}f^{(k+1)} + \ldots + c_m f^{(m)}|$$

$$\leq \max_{\ell=1,\ldots,m-k} \frac{|c_{k+\ell}|}{r^{\ell}} M(r,f^{(k)}) < |c_k| M(r,f^{(k)})$$

for all large r. Thus

$$M(r,Lf) = |c_k| M(r,f^{(k)}) = cM(r,f^{(k)}) \tag{2.7}$$

for all large r. According to Lemma 2.2 we have for $f = \sum a_n x^n$

$$\begin{aligned} M(r,f^{(k)}) &= \max_n |n(n-1) \ldots (n-k+1)a_n| r^{n-k} \\ &\geq \max_n \frac{|a_n| r^n}{(nr)^k} \geq \frac{M(r,f)}{(r \frac{d\mu}{d\rho})^k} . \end{aligned} \tag{2.8}$$

Combining (2.5), (2.7) and (2.8) we get, for r large,

$$r \frac{d\mu}{d\rho} \geq \left(\frac{cM}{\phi(M)}\right)^{1/k} \tag{2.9}$$

or

$$\frac{r^2 \phi(M)^{1/k}}{M^{1+1/k}} dM \geq c^{1/k} dr = c' dr . \tag{2.10}$$

Since f is transcendental we have $r^2 < M(r,f)^{\delta}$ for any $\delta > 0$ and all large r. Thus (2.10) becomes

$$\frac{\phi(M)^{1/k}}{M^{1+1/k-\delta}} dM \geq c' dr \tag{2.11}$$

for large r. Since the integral of the right side diverges the integral of the left side must diverge also.

PROOF OF THEOREM 1.1 FOR ENTIRE FUNCTIONS. Assume that f is an entire function in $\mathcal{R}$. Since $\mathcal{R}$ is finite dimensional over $\mathcal{L}$ the function f must satisfy a differential equation of the form

$$L(f^n) = L_1 f + L_2(f^2) + \dots + L_{n-1}(f^{n-1})$$

where $L \neq 0$ and $L, L_1, \dots, L_{n-1} \in \mathcal{L}$. Thus

$$\begin{aligned} M(r, L(f^n)) &= M(r, L_1 f + \dots + L_{n-1}(f^{n-1})) \leqslant cM(r,f)^{n-1} \\ &= cM(r, f^n)^{1-1/n} \end{aligned} \tag{2.12}$$

for all large r. Inequality (2.12) has the form of (2.5) with $\phi(t) = ct^{1-1/n}$. However for this ϕ we have

$$\int^{\infty} \frac{\phi(M)^{1/k}}{M^{1+1/k-\delta}}\, dM < \infty$$

for all k and $\delta < 1/kn$ in contradiction to Lemma 2.4, unless f is a polynomial.

The rest of the proof of Theorem 1.1 is essentially the same as in the case of analytic functions of a complex variable [1]:

PROOF OF THEOREM 1.1 FOR RINGS CONTAINING NON CONSTANT ENTIRE FUNCTIONS. Since we may assume without loss of generality that $\mathcal{R}$ contains the constant functions; it follows that if

R contains any nonconstant entire function, that is a nonconstant polynomial, we must have $x \in R$. Now let f be a nonentire element of R. If f has a finite set of poles $p_1,\dots,p_n$ with multiplicities $m_1,\dots,m_n$ then $(x-p_1)^{m_1}\dots(x-p_n)^{m_n} f$ is an entire element of R and hence a polynomial. Thus f is rational. If f has an infinite set of poles $\{p_1,p_2,\dots\}$ with multiplicities $\{m_1,m_2,\dots\}$ then the functions $f_0 = f$, $f_1 = (x-p_1)^{m_1} f_0$, $f_2 = (x-p_2)^{m_2} f_1,\dots$ form an infinite set of functions linearly independent over L since in any linear combination

$$L_k f_k + L_{k+1} f_{k+1} + \dots + L_n f_n, \quad L_k \neq 0$$

only the first term has a pole at p_k. Thus all elements of R are rational. If p is the pole of an element of R then $1/(x-p)$ R. Clearly for different $p_1,\dots,p_n$ the functions $1/(x-p_1)$, $\dots,1/(x-p_n)$ are linearly independent over L and thus the union of the set of poles of the functions of R is finite, equal to the dimension of R over L.

<u>LEMMA 2.13</u> Let R satisfy the hypothesis of Theorem 1.1 and contain nonentire meromorphic functions. Then R contains a meromorphic function with a minimal set of poles.

<u>Proof</u>. If the lemma were false then there would exist functions $f_1,f_2,\dots$ in R whose poles form a properly decreasing sequence of sets. Say

p_n is a pole of f_n but not of f_{n+1}. Then $L_k f_k + L_{k+1} f_{k+1} + \dots + L_m f_m$ with $L_k \neq 0$ must have a pole at p_k and thus $\mathcal{R}$ contains an infinite sequence of functions linearly independent over $\mathcal{L}$.

LEMMA 2.14 Let $\mathcal{R}$ be a ring of meromorphic functions on $\mathcal{D}$ which is finite dimensional over $\mathcal{L}$ and contains no nonconstant entire functions. Let $f \in \mathcal{R}$ be a function with minimal pole set $\mathcal{P} = \{p_1, \dots\}$ and pole of minimal order at p_1. Then all functions $g \in \mathcal{R}$ with pole set $\mathcal{P}$ have the form $g = Lf + c$; $L \in \mathcal{L}$, $c \in \mathcal{D}$.

Proof. Let

$$f = \frac{a_{-k}}{(x-p_1)^k} + \frac{a_{-k+1}}{(x-p_1)^{k-1}} + \dots$$

and

$$g = \frac{b_{-\ell}}{(x-p_1)^\ell} + \frac{b_{-\ell+1}}{(x-p_1)^{\ell-1}} + \dots, \quad \ell \geqslant k .$$

Then

$$g_1 = g - \frac{b_{-\ell}}{a_{-k}} \frac{1}{(-k)(-k-1)\dots(-\ell+1)} D^{\ell-k} f$$

has a pole or order $\leqslant \ell - 1$ at p_1. Repeating this process we get a linear operator $L \in \mathcal{L}$ so that $g - Lf$ has a pole of order less than k at p_1. By the minimality of k it follows that $g - Lf$ is regular at p_1 and by the minimality of $\mathcal{P}$ it follows that $g - Lf$ is entire and hence constant.

LEMMA 2.15 Let $\mathcal{R}$ satisfy the hypotheses of Lemma 2.14; then a minimal pole set consists of a single point.

Proof. Assume that a minimal pole set, $\mathcal{P}$, contains two points p_1, p_2 and the function f with pole set $\mathcal{P}$ has a pole of minimal order at p_1. According to Lemma 2.14 there exists a linear differential operator

$$L = c_m D^m + c_{m-1} D^{m-1} \cdot \ldots + c_1 D + c_0$$

and a constant c so that $f^2 = Lf + c$. Thus f has a pole of order at both p_1 and p_2. Set

$$f = \sum_{n=-m}^{\infty} a_{n1}(x-p_1)^n = \sum_{n=-m}^{\infty} a_{n2}(x-p_2)^n .$$

We wish to prove that $a_{n1} = a_{n2}$ for all n, leading to the identity $f(x) = f(x+p_2-p_1)$ which cannot hold for a nonconstant meromorphic function.

Comparing coefficients of $(x-p_i)^{-m}$ we get

$$a_{-mi}^2 = c_m(-m)(-m - 1) \ldots (-2m + 1) a_{-mi}$$

so that

$$a_{-mi} = (-1)^m \frac{(2m-1)!}{(m-1)!} c_m = a_{-m}; \quad i = 1,2$$

is independent of i. Now assume that $a_{ki} = a_k$ is independent of i for $k < n$; then comparing coefficients of $(x-p_i)^{n-m}$ in f^2 and $Lf + c$ we get

$$(2.16)\quad 2a_{-m}a_{ni} + \ldots = c_m n(n-1)\ldots(n-m+1)a_{ni} + \ldots$$

where the terms not written are independent of i. Thus (2.16) implies that a_{ni} is independent of i unless

$$2a_{-m} = \frac{n!}{(n-m)!} c_m ,$$

that is

$$(-1)^m 2 \frac{(2m-1)!}{(m-1)!} = \frac{n!}{(n-m)!}$$

which can happen only if m is even and $n = 2m$. In that case we use the differential equation

$$(2.17)\qquad f^3 = L'f + c'$$

with

$$L' = c'_{2m}D^{2m} + \ldots ;$$

comparing coefficients of $(x-p_1)^{-3m}$ in (2.17) we get

$$a^3_{-m} = \frac{(3m-1)!}{(m-1)!} c'_{2m}a_{-m}$$

or

$$(2.18)\qquad c'_{2m} = \frac{(m-1)!}{(3m-1)!} a^2_{-m} .$$

Now comparing constant terms in (2.17) we get

$$(2.19)\quad 3a^2_{-m}a_{2mi} + \ldots = c'_{2m}(2m)!\, a_{2mi} + \ldots$$

where the terms not written are independent of i.

The coefficients on the sides of (2.19) are

$$3 > \frac{(m-1)!(2m)!}{(3m-1)!}$$

so that $a_{2mi} = a_{2m}$ is independent of i.

LEMMA 2.20 Let $\mathcal{R}$ satisfy the hypotheses of Lemma 2.14 and let $f \in \mathcal{R}$ have a single pole of minimal order. Then f is rational.

Proof. According to Lemma 2.14 we have

$$(2.21) \qquad f^2 = Lf + c, \quad L \in \mathcal{L}, \quad c \in \mathcal{D} .$$

Let p be the pole of f and write

$$f(x) = \frac{a_{-m}}{(x-p)^m} + \ldots + \frac{a_{-1}}{x-p} + g(x) = R(x) + g(x)$$

where g is entire. Thus (2.21) becomes

$$(2.22) \qquad g^2 = -2Rg + Lg + c + LR-R^2 .$$

From (2.22) it follows that

$$M(r,g)^2 = M(r,g^2) = O(M(r,g)) + O(1)$$

for large r, which is impossible unless g is constant.

PROOF OF THEOREM 1.1 FOR RINGS WITHOUT NON-CONSTANT ENTIRE FUNCTIONS. According to Lemma 2.20 we know that $\mathcal{R}$ contains at least one rational function f with a single pole p. Now let $g \in \mathcal{R}$ have a pole set $P = \{p_1, p_2, \ldots\}$ not counting the point p if it should happen to be a pole cf g. Let the multiplicity of the pole p_i be m_i.

If P is infinite then the functions $g_1(x) = (f(x) - f(p_1))^{m_1} g(x)$, $g_2(x) = (f(x)-f(p_2))^{m_2} g_1(x), \ldots$ have a properly decreasing sequence of pole sets and hence are linearly independent over L. Thus $P = \{p_1, p_2, \ldots, p_n\}$ is finite and

$$g_0(x) = g(x) \prod_{i=1}^{n} (f(x) - f(p_i))^{m_i}$$

has poles only at p. Thus by Lemma 2.14 we have $g_0(x) = L_0 f + c$ is rational and hence $g(x)$ is rational.

Let $f_1, \ldots, f_n$ be a maximal set of functions in R which are linearly independent over L. Then for every $g \in R$ there is a nonzero $L \in L$ so that $Lg = L_1 f_1 + \ldots + L_n f_n$ and hence the poles of g are in the finite union of the pole sets of $f_1, \ldots, f_n$.

3. On the Growth Rate of Differential Expressions of an Entire Function.

In this section we show that the Main Lemma 2.4 of the preceding section does not hold if we permit linear differential operators with polynomial coefficients and restrict attention to entire functions of a p-adic variable. Thus in this section we assume that the domain $D = \Omega_p$ is the completion of the algebraic closure of Q_p, the field of p-adic numbers.

THEOREM 3.1 Given any positive monotonic function $\phi(t)$ which increases to infinity with t, however slowly, there exists a constant $c \in \Omega_p$ and a transcendental entire function f on Ω_p so that

$$(3.2) \qquad M(r,(xD-c)f) < \phi(M(r,f))$$

for all $r \geqslant 0$.

Proof. We wish to construct a p-adic Liouville number $c = \lim n_i$ where the n_i form a rapidly converging sequence of positive integers. The corresponding entire function

$$f(x) = \prod_{i=1}^{\infty} p^{in_i} x^{n_i}$$

satisfies

$$(3.4) \qquad g(x) = (xD-c)f(x) = \prod_{i=1}^{\infty} p^{in_i}(n_i-c)x^{n_i} .$$

It remains to determine the sequence n_i so that the inequality (3.2) is satisfied.

The maximum modulus diagram (Definition 2.1) $\mu(\rho)$ has edges with slopes n_i for $\rho_{i-1} \leqslant \rho \leqslant \rho_i$ where $\rho_0 = -\infty$ and

$$\rho_i = i + \frac{n_{i+1}}{n_{i+1} - n_i} \log p; \quad i = 1,2,\dots .$$

These values are determined by making consecutive terms in (3.3) have the same absolute value. If we assume $n_{i+1} > 2n_i$ we get

$$(3.5) \qquad (i + 1)\log p < \rho_i < (i + 2)\log p .$$

If we assume that $|c - n_i|$ is monotonically

decreasing then for $r \leq r_i = e^{\rho_i}$ we get

$$M(r,g) \leq \max_{j \leq i} |c - n_j| p^{-jn_j} r^{n_j} .$$

It thus suffices to prove that

$$|c - n_i| p^{-in_i} r^{n_i} \leq \phi(M(r,f))$$

for all $r \geq r_{i-1}$, that is

$$|c - n_i| p^{-in_i} r^{n_i} \leq \phi(p^{-jn_j} r^{n_j}) \tag{3.6}$$

for $r_{j-1} \leq r \leq r_j$; $j = i, i + 1, \ldots$. For $j \leq i+2$ we could satisfy (3.6) by

$$|c-n_i| \leq r_{i+2}^{-n_i} \phi(0)$$

which according to (3.5) would be satisfied if

$$|c - n_i| \leq \phi(0) p^{-(i+4)n_i} . \tag{3.7}$$

For $j > i + 2$ we have

$$p^{-jn_j} r_{j-1}^{n_j} = p^{n_j n_{j-1}/(n_j - n_{j-1})} > p^{n_{j-1}} .$$

Thus (3.6) is satisfied if we choose n_{j-1} so large that

$$|c - n_i| p^{-in_i} r_j^{n_i} < r_j^{n_i} < p^{(j+2)n_i}$$

$$\leq p^{(j+2)n_{j-2}} < \phi(p^{n_{j-1}}) .$$

Which is certainly true if we assume

$$(3.8)\qquad n_j > \log_p \phi^{-1}(p^{(j+3)n_{j-1}}); \quad j = 2,3,\ldots .$$

To satisfy (3.7) and (3.8) simultaneously we choose $n_i = p^{\nu_i}$ with (3.8) satisfied and (3.7) being a consequence of $|c - n_i| = 1/n_{i+1}$ where

$$(3.9)\qquad n_j > \frac{1}{\phi(0)} p^{(j+3)n_{j-1}} .$$

We clearly can construct a sequence of integers ν_j recursively so that both (3.8) and (3.9) are satisfied.

REFERENCES

[1] E.G. Straus, Differential rings of meromorphic functions, to appear in Acta Arithmetica.

University of California, Los Angeles, California

NEW RESULTS AND APPLICATIONS OF THE THEORY OF DIOPHANTINE APPROXIMATIONS

Paul Turán

1. The results and applications refer to the trend in the theory of diophantine approximations which formulates its problems in "operatortype sums"

$$\sum_{j=1}^{n} \phi_j(\nu) z_j^{\nu} , \quad \sum_{j=1}^{n} \phi_j(z_j) z_j^{\nu}$$

where the ϕ_j's are polynomials, ν being an integer and the z_j's fixed complex numbers. In this talk we shall be exclusively concerned with the case when the ϕ_j's are complex constants b_j , i.e. we are dealing with generalized power-sums

$$g(\nu) = \sum_{j=1}^{n} b_j z_j^{\nu} . \tag{1.1}$$

The problems are of minimax type where in most cases good estimations of the value of minimax is enough for the applications. The applicability of these theorems is due to the fact that, whereas for a sum

$$|\eta_1 + \eta_2 + \ldots + \eta_n|$$

η_j comples numbers

in general no <u>lower</u> bound can be given, apart from the obvious case, knowing something on the <u>terms</u>, say

$$\min_j |z_j| \quad \text{or} \quad \max_j |z_j| \quad \text{or} \quad \Big(\sum_j |z_j|^2\Big)^{\frac{1}{2}}$$

they <u>can</u> produce such lower bounds which are adaptable to very many situations.

Having a limited time I shall not dwell much upon why this trend belongs to the theory of diophantine approximations; the facts that Dirichlet's and Kronecker's classical theorems can be proved, even in stronger forms, within this circle of problems and ν is an integer in the results, are enough indications. This last fact is particularly important in several applications whereas it was never essential in the applications of H. Bohr who first applied diophantine approximation to analysis.

2. I shall start with a slight digression. In the trend in question the various upper and lower estimations of generalized power-sums are <u>aims</u> for the theory and <u>tools</u> for applications. As tools of the theory they were used since H. Weyl's theory of uniform distribution mod 1. The main result of this theory says, in a form given by P. Erdős and myself in 1948, that if $a_1, a_2, \ldots$ is

a real sequence and

$$\psi(\nu,n) = \sum_{j=1}^{n} (e^{2\pi i a_j})^{\nu}$$

then for $0 \leq \alpha < \beta \leq 2\pi$ and integer $m > 0$ the inequality

$$\Big| \sum_{\substack{\nu \leq n \\ \alpha \leq a_\nu \leq \beta \bmod 1}} 1 - (\beta-\alpha)n \Big| < C\Big(\frac{n}{m+1} + \sum_{\nu=1}^{m} \frac{1}{\nu}|\psi(\nu,n)|\Big)$$

holds with a positive numerical constant C.

Perhaps it is less known that this theorem from the classical theory of diophantine approximations, generalized to more dimensions by Koksma and Szüsz, has an application to the deterministic kinetic theory of gases as we found with Egerváry in 1951, fulfilling a desideratum of H. Steinhaus. We could namely deduce in the case when the particles are in a parallelepiped that they uniformly distributed in all "decent" domains apart from "small"time, which was previously a postulate. The initial position of the particles proved to be unessential but we made a rather artificial assumption on the initial velocities. This was quite essentially generalized in the last few years by Hlawka; his papers on this subject are in my opinion much less known than they deserve.

An essential progress could be made having an insight into the orbit of a point moving on a billiard table of the shape of an isoceles rectangular triangle under the law of reflection.

3. Coming back to my proper subject, in a course in Ann Arbor last year I centered tha applications of this theory around four main theorems. Today I shall speak only on three of them, review on some developments of them I obtained while writing out a final form of a book on it (which is not final any more) and discuss a few new applications.

The first main theorem asserted that at fixed complex b_j numbers and integer $m \geq 0$, variable $z_1, \ldots, z_n$ with

$$1 = |z_1| \leq |z_2| \leq \ldots \leq |z_n| \tag{3.1}$$

the inequality

$$\min_{z_j} \max_{\substack{m+1 \leq \nu \leq m+n \\ \nu \text{ integer}}} |g(\nu)| \geq \left(\frac{n}{2e(m+n)}\right)^n |g(0)| \tag{3.2}$$

holds. Its proof is short and elementary-algebraic. Suitable substitution gives at once that if a, d are positive, $\alpha_1, \ldots, \alpha_n$, $b_1, b_2, \ldots, b_n$ complex numbers, Λ is defined by

$$\min_{j=1,\ldots,n} \operatorname{Re} \alpha_j = \Lambda \tag{3.3}$$

and

$$f(t) = \sum_{j=1}^{n} b_j e^{\alpha_j t} , \tag{3.4}$$

then the inequality

$$(3.5)\qquad |f(0)| \leq \left(2e\,\frac{a+d}{d}\right)^n \max_{a\leq t\leq a+d} |f(t)|e^{-\Lambda t}$$

holds. $f(t)$ is the general solution of the differential equation

$$(3.6)\qquad h^{(n)} + a_1 y^{(n-1)} + \ldots + a_n y = 0$$

with constant coefficients if only all zeros of the equation

$$(3.7)\qquad x^n + a_1 x^{n-1} + \ldots + a_n = 0$$

are simple. Trivial passage to limit gives at once that (3.5) holds without any restriction on the coefficients of (3.6) too, for all solutions of (3.6). L_2 - forms of (3.5) could be deduced from (3.2) too, the strongest being due to Tijdeman and asserts that for $a > 0$ the inequality

$$(3.8)\qquad \int_0^d |f(t)|^2 e^{-2\Lambda t}dt \leq \left(2e\,\frac{a+d}{d}\right)^{2n} \int_a^{a+d} |f(t)|^2 e^{-2\Lambda t}dt$$

holds for all solutions of the equation (3.6).

4. Combining (3.8) with a perturbation process one can deduce the following theorem.

Let $A = (a_{ik})$ an $n \times n$ matrix with eigenvalues $\lambda_1, \lambda_2, \ldots, \lambda_n$ so that

$$(4.1)\qquad \max_{i,k}|a_{ik}| \leqslant M \qquad \min_j \operatorname{Re}\lambda_j = \Lambda .$$

Let $X(t)$ be an $n \times 1$ columnvector which satisfies the equation

$$(4.2)\qquad \frac{dX(t)}{dt} = AX(t) + W(X,t)$$

where the $n \times 1$ columnvector $W(Z,t)$ satisfies for all $n \times 1$ Z-columnvectors for $t \geqslant 0$ the inequality

$$(4.3)\qquad \frac{|W(Z,t)|}{|Z|} \leqslant e^{-4n-\delta(nM+2|\Lambda|)}$$

with a constant $\delta > 0$. Then we have for all $\beta > 0$ the inequality

$$(4.4)\qquad \int_0^{\frac{\delta}{2}} |X(t)|^2 e^{-2\Lambda t}dt \leqslant e^{\frac{20}{\delta}\beta} \int_\beta^{\beta+\delta} |X(t)|^2 e^{-2\Lambda t}dt .$$

(3.5) and (3.8) offer other ways to obtain general theorems of the above type for nonlinear differential equations.

5. Another recent application of (3.2) refers to function theory. Let $r_n(z)$ a rational function of nth degree; more exactly let

$$r_n(z) = \frac{u(z)}{v(z)} \tag{5.1}$$
$$0 \leq \deg u(z) < \deg v(z) = n .$$

Then the inequality of (3.2) permits to give a nontrivial lower bound of the absolute square integral of $r_n(z)$ along a circle in dependence on $n, |r(0)|$, and the maximal distance of the poles from the center z_o of the circle. We may suppose without loss of generality $z_o = 0, |r_n(0)| = 1$ and the poles $\xi_1, \xi_2, \ldots, \xi_n$ with

$$(0<)|\xi_1| \leq |\xi_2| \leq \ldots \leq |\xi_n| = 1 . \tag{5.2}$$

Then we found with Rahman via (3.2) the inequality

$$\frac{1}{2\pi} \int_{-\pi}^{\pi} |r_n(\rho e^{i\phi})|^2 d\phi \geq 1 + \frac{\rho^{2n}}{\sum_{\nu=0}^{n-1} \binom{n}{\nu}^2 \rho^{2\nu}} \tag{5.3}$$

for $0 < \rho < |\xi_1|$. As the example $r_n(z) = \frac{1}{1-z^n}$ shows, (5.3) is for small ρ's essentially best-possible. The interest of this theorem lies in the complete independence of the lower bound on the coefficients of $u(z)$, in the weak dependence on those of $v(z)$. For polynomials we can have no such lower bound apart from the trivial 1.

6. Leaving aside further new applications of (3.2) to the theory of differential-difference equations we shall turn to some new improvements

of the second main theorem. Whereas the first main theorem (3.2) estimates $|g(\nu)|$ from below for appropriate "dense" values of integer ν's "essentially" by its minimal term, the second one is doing the same "essentially" by its maximal term. More exactly this says in its original form that if m is a nonnegative integer, $b_1, b_2, \ldots b_n$ are fixed, $z_1, z_2, \ldots, z_n$ variable complex numbers with

$$(6.1) \qquad 1 = |z_1| \geq |z_2| \geq \ldots \geq |z_n|$$

then the inequality

$$(6.2) \qquad \min_{z_j} \max_{\substack{m+1 \leq \nu \leq m+n \\ \nu \text{ integer}}} |g(\nu)| \geqq \left(\frac{n}{8e(m+n)}\right)^n \min_{j=1,\ldots,n} |b_1 + \ldots + b_j| .$$

holds. The several applications of this theorem in analytical number theory refer mostly to the cases when $Rb_j \geq 0$ or even

$$(6.3) \qquad b_1 = \ldots = b_n = 1 \;;$$

trivial examples show that the dependence of the right side of (6.2) on the b_j's cannot be better, say

$$|b_1 + \ldots + b_n| \quad \text{or} \quad \min_j |b_j| .$$

However, it was observed by my late collaborator S. Knapowski that by a slight modification of the proof the range of the index j in (6.2) could be narrowed and even this improvement enlarged considerably the range of applications. Recently, I observed that (6.2) can be given in an apparently much more applicable form. Namely to given integer $m \geq 0$, complex b_j's and z_j's satisfying (6.1) there is an integer ν_0 with

$$m + 1 \leq \nu_0 \leq m + n$$

such that

$$(6.4) \qquad |g(\nu_0)| \geq \left(\frac{n}{16e(m+n)}\right)^n \min_{L_1 \leq j \leq L_2} |b_1 + \ldots + b_j|$$

where L_1 is any index with

$$(6.5) \qquad |z_{L_1}| \geq 1 - \frac{n}{2(m+n)}$$

(such an index always exists) and L_2 is any index with

$$(6.6) \qquad |z_{L_2}| \leq \frac{m}{m+n}$$

if it exists; if not, take $L_2 = n$. In <u>some</u> applications it does not matter much that in (6.4) the constant $8e$ is replaced by $16e$; in some others however it would be very important to replace the constant $8e$ in (6.2) by $(1+\varepsilon)$ for $m > m_0(\varepsilon)$ for <u>some</u> b_j-systems, in particular in the important case (6.3).

7. In the applications of all the three theorems mentioned in 6. we had to cope with <u>partial-sums</u> $\sum_{j=1}^{j} b_{\ell}$. In some cases however the first few b's (their order being determined by (6.1)) are small which makes the application of (6.4) useless. In such cases one can hope for success only if we have an analogous theorem with <u>part-sums</u>

$$\sum_{\ell=j_1}^{j_2} b_j$$

(again under (6.1)). I really found recently such a theorem. Let

$$(7.1) \qquad 1 \leqslant k \leqslant n \leqslant m$$

integers and we define the indices ℓ_1 and ℓ_2 dropping the normalization $\max_j |z_j| = 1$ by

$$(7.2) \qquad \begin{aligned} |z_1| \geqslant |z_2| \geqslant \ldots \geqslant |z_{\ell_1 - 1}| > \frac{m+4n}{m+2n} |z_k| \geqslant |z_{\ell_1}| \\ \geqslant \ldots \geqslant |z_k| \geqslant \ldots \geqslant |z_{\ell_2}| \geqslant \frac{m}{m+2n} |z_k| \\ > |z_{\ell_2+1}| \geqslant \ldots \geqslant |z_n| \end{aligned}$$

further the indices L_1, L_2 be such that

$$(7.3) \qquad 1 \leqslant L_1 \leqslant \ell_1 \leqslant \ell_2 \leqslant L_2 \leqslant n$$

Then there is an integer ν_o such that

$$(7.4) \qquad m + 1 \leqslant \nu_o \leqslant m + n$$

and

$$(7.5)\qquad |g(\nu_o)| \geq \left(\frac{n}{12e(m+2n)}\right)^{2n} |z_k|^{\nu_o} \min|b_{r_1}+\ldots+b_{r_2}|$$

where the min refers to all (r_1,r_2) pairs with

$$(7.6)\qquad L_1 \leq r_1 \leq r_2 \leq L_2 .$$

The proof of this theorem is more difficult than that of (6.2). Essential use is made of the theorem, conjectured previously by Mahler, that

$$(7.7)\qquad \max_{-1\leq x\leq +1} |\pi_\kappa(x)| \geq \frac{1}{2^{\kappa-1}} \prod_j{}' |\xi_j|$$

for all polynomials

$$\pi_\kappa(z) = z^\kappa + \ldots + c_o$$

of degree κ where the product in (7.7) is to be extended to all zeros ξ_j of $\pi_\kappa(z)$ with $|\xi_j| > 1$ (empty product being 1).

A notable case arises when m is so big that instead of (7.2) we have

$$(7.8)\qquad \begin{aligned} &|z_1| \geq |z_2| \geq \ldots \geq |z_{k-1}| > \frac{m+4n}{m+2n}|z_k| > |z_k| \\ &> \frac{m}{m+2n}|z_k| > |z_{k+1}| \geq \ldots \geq |z_n| . \end{aligned}$$

Then (7.4)-(7.5) takes the simple form that for a suitable integer ν_o with

$$(7.9)\qquad m+1 \leq \nu_o \leq m+n$$

the inequality

$$(7.10) \qquad |g(\nu_o)| \geq \left(\frac{n}{2e(m+2n)}\right)^{2n} |b_k z_k^{\nu_o}|$$

holds, which means indeed something stronger than a lower bound of $|g(\nu_o)|$ by its <u>exact</u> minimal term for suitable ν_o. No doubt the inequalities (7.5) and (7.10) enlarge considerably the adaptability of the second main theorem to various situations.

8. Next we turn to the fourth main theorem, which is a "one-sided" theorem. With the notations (1.1) and (6.1) let the b_j's fixed complex numbers such that all partial sums

$$(8.1) \qquad \sum_{\ell=1}^{j} \mathrm{Re}\, b_\ell$$

are all of the same sign and m be a given non-negative integer. The z_j's should be complex variables which are restricted beside (6.1) also by the argument condition

$$(8.2) \qquad \kappa \leq |\mathrm{arc}\, z_j| \leq \pi\,, \qquad j = 1,2,\ldots,n$$

where κ is a fixed positive number with

$$(8.3) \qquad 0 < \kappa \leq \frac{\pi}{2}\,.$$

Then the theorem in question asserts the existence of integer ν_1 and ν_2 with

$$(8.4) \qquad m + 1 \leq \nu_1, \qquad \nu_2 \leq m + n(3+\frac{\pi}{\kappa})$$

so that the inequalities

$$(8.5)\quad \mathrm{Re}\, g(\nu_1) \geqslant \frac{1}{3n}\left(\frac{n}{8e(m+n(3+\frac{\pi}{\kappa}))}\right)^{2n} \min_j \left|\sum_{\ell=1}^{j} \mathrm{Re}\, b_\ell\right|$$

and

$$(8.6)\quad \mathrm{Re}\, g(\nu_2) \leqslant -\frac{1}{3n}\left(\frac{n}{8e(m+n(3+\frac{\pi}{\kappa}))}\right)^{2n} \min_j \left|\sum_{\ell=1}^{j} \mathrm{Re}\, b_\ell\right|$$

hold. Trivial examples show that the restriction (8.2) is necessary that <u>both</u> (8.5) and (8.6) hold.

9. This theorem had so far several applications in the analytical theory of numbers starting with the theorem of Knapowski according which the number of sign changes of $\{\pi(x) - \int_2^x \frac{dv}{\log v}\}$ for $0 < x < T$ is greater than const. log log log log T, in still greater contrast to Riemann's well-known contention than Littlewood's famous original disproof which furnished merely an infinity of sign changes ($\pi(x)$ as usual stands for the number of primes $\leqslant x$). Again the dependence of (8.5) and (8.6) on the b_ℓ's is what makes difficulties in applications of them. Recently I found that the above theorem can be refined analogously as the second main theorem in 7. More exactly let us suppose again (7.1) and the argument-condition (8.2) and putting

$$(9.1)\qquad \frac{4n}{m+n(t+\frac{\pi}{\kappa})} = \Delta$$

we define (again without the normalization $\max_j |z_j| = 1$) the indices ℓ_1 and ℓ_2 this time by

$$|z_1| \geq |z_2| \geq \dots \geq |z_{\ell_1-1}| > (1+\Delta)|z_k| \geq |z_{\ell_1}|$$

(9.2)

$$\geq \dots \geq |z_k| \geq \dots \geq |z_{\ell_2}| \geq (1-\Delta)|z_k|$$

$$> |z_{\ell_2+1}| \geq \dots \geq |z_n| .$$

Suppose further that for some L_1, L_2 integers with

$$1 \leq L_1 \leq \ell_1 \leq \ell_2 \leq L_2 \leq n \tag{9.3}$$

all sums

$$\sum_{r_1 \leq \ell \leq r_2} \operatorname{Re} b_\ell \tag{9.4}$$

are of the same sign for

$$L_1 \leq r_1 \leq r_2 \leq L_2 . \tag{9.5}$$

Then there are integers ν_1 and ν_2 with

$$m + 1 \leq \nu_1, \qquad \nu_2 \leq m + n(3+\tfrac{\pi}{\kappa}) \tag{9.6}$$

such that

$$\operatorname{Re} g(\nu_1) \geq \frac{|z_k|^{\nu_1}}{3n}\left(\frac{n}{5e(m+n(5+\frac{\pi}{\kappa}))}\right)^{4n} \min\left|\operatorname{Re}\sum_{\ell=r_1}^{r_2} b_\ell\right| \tag{9.7}$$

and

$$\operatorname{Re} g(\nu_2) \leq -\frac{|z_k|^{\nu_2}}{3n}\left(\frac{n}{5e(m+n(5+\frac{\pi}{\kappa}))}\right)^{4n} \min\left|\operatorname{Re}\sum_{\ell=r_1}^{r_2} b_\ell\right| \tag{9.8}$$

where the min refers to all (r_1, r_2) pairs in

(9.5).

The proof of this theorem is also rather difficult.

10. Again a notable special case arises when m is so big that instead of (9.2) we have

$$|z_1| \geq |z_2| \geq \dots \geq |z_{k-1}| > (1+\Delta)|z_k| > |z_k|$$

$$(10.1) \quad > (1-\Delta)|z_k| > |z_{k+1}| \geq \dots \geq |z_n| .$$

Then (9.6)-(9.7)-(9.8) gives that for suitable integers ν_1 and ν_2 with

$$(10.2) \quad m + 1 \leq \nu_1, \qquad \nu_2 \leq m + n(3+\frac{\pi}{\kappa})$$

the inequalities

$$(10.3) \quad \operatorname{Re} g(\nu_1) \geq \frac{|z_k|^{\nu_1}}{3n}\left(\frac{n}{5e(m+n(5+\frac{\pi}{\kappa}))}\right)^{4n} |\operatorname{Re} b_k|$$

$$(10.4) \quad \operatorname{Re} g(\nu_2) \leq -\frac{|z_k|^{\nu_2}}{3n}\left(\frac{n}{5e(m+n(5+\frac{\pi}{\kappa}))}\right)^{4n} |\operatorname{Re} b_k|$$

hold.

No doubt that the inequalities (9.7)-(9.8) and (10.3)-(10.4) will play an essential role in the comparative prime number theory.

11. Finally we shall indicate a new application of the one sided theorem (8.5)-(8.6). This refers to the distribution of the p primes in various residue classes mok k. Cebysev conjec-

tured more than a hundred years ago that

$$(11.1)\qquad \lim_{\omega\to\infty} \sum_{p>2} (-1)^{\frac{p-1}{2}} e^{-\frac{p}{\omega}} = -\infty$$

which says "in Abelian sense" that there are more primes $\equiv 3 \bmod 4$ than $\equiv 1 \bmod 4$. If $s = \sigma + it$, then it was shown in 1918 by Hardy-Littlewood and Landau that (11.1) is equivalent with the assertion

$$(11.2)\qquad \sum_{n=0}^{\infty} \frac{(-1)^n}{(2n+1)^s} \neq 0 \qquad \text{for} \qquad \sigma > \frac{1}{2}$$

which is not yet decided. That Cebysev's assertion's direct form

$$\pi(x,4,1) \leq \pi(x,4,3)$$

is false for suitable $x_1 < x_2 < \ldots \to \infty$, was shown by Littlewood in 1914; $\pi(x,k,\ell)$ stands as usual for the number of primes $\equiv \ell \bmod k$ and not exceeding x. Now one of the main problems of the comparative prime number theory consists in showing that each functions

$$(11.3)\qquad \pi(x,k,\ell_1) - \pi(x,k,\ell_2) \qquad 1 \leq \ell_1 \neq \ell_2 < k$$

changes sign infinitely often and to find an upper bound $A(k)$ such that each function in (11.3) changes sign on the segment $(2, A(k))$. Various results have been achieved towards the first part of the problem, mostly ineffective ones; they depend as well as the effective ones (which *do* give explicit upper bounds for $A(k)$), on various assumptions on the "small" zeros of Dirichlet's $L(s,k,\chi)$ functions. The weakest of them is that there is a $B = B(k) > 0$ such that no $L(s,k,\chi)$ functions vanishes for

$$(11.4) \qquad \sigma \geq \frac{1}{2}, \qquad |t| \leq B(k),$$

which can be called Haselgrove condition. Particularly difficult are the problems where *some* $\pi(x,k,\ell_1)$'s with quadratic residue ℓ_1's are compared to equally many $\pi(x,k,\ell_2)$'s with quadratic nonresidue ℓ_2's, notably when *one* $\pi(x,k,\ell_1)$ with $\ell_1 \not\equiv 1 \bmod k$ is compared with *one* $\pi(x,k,\ell_2)$, the last ones being expected to preponderate in some sense. As a new application of theorem (8.5)-(8.6) one can prove that putting

$$(11.5) \qquad \sum_{\substack{p \leq x \\ p \equiv \ell \bmod k}} \log p = \Theta(x,k,\ell)$$

and supposing the truth of the Haselgrove condition (11.4) then for an *explicitly calculable*

$$(11.6) \qquad c = c(k,B(k))$$

and $T > c$ the interval

$$(11.7) \qquad \exp(\log_3 T)^{1/300} \leq x \leq T$$

contains for each quadratic nonresidue ℓ an x_0 for which the inequality

(11.8) $\frac{3}{4}\,\Theta(x_o,k,1) + \frac{1}{4}\,\Theta(x_o,k,\ell^2) > \Theta(x_o,k,\ell)$

even

$$\frac{3}{4}\,\Theta(x_o,k,1) + \frac{1}{4}\,\Theta(x_o,k,\ell^2)$$
$$> \Theta(x_o,k,\ell) + \sqrt{x_o}\;e^{-\sqrt{\log x_o}}$$

holds, against the above mentioned expectations. The occuring constants are explained by the fact that

$$3 - 4\cos\phi + \cos 2\phi$$

is a nonnegative trigonometrical polynomial. It is quite possible that in (11.8) $\Theta(x,k,\ell)$ can be replaced by $\pi(x,k,\ell)$.

Mathematical Institute of the Hungarian Academy of Sciences, Budapest, Hungary

GOOD LATTICE POINTS MODULO PRIMES AND COMPOSITE NUMBERS

S.K. Zaremba

The need for good lattice points, to use an expression introduced by Hlawka [1], arises in connection with the numerical integration of functions over multidimensional intervals, which, of course, can be reduced to unit intervals in s dimensions,

$$Q^s: \quad 0 \leq x_i < 1 \quad (i=1,\ldots,s) .$$

The requirement that the domain of integration should be a multidimensional interval or a domain easily reducible to such an interval is undoubtedly a very serious limitation. However, the problem becomes incomparably more complicated when we pass to more general domains of integration, for instance to convex domains, and this case will not be treated here.

Vectors will always be denoted by underlined letters; their coordinates will be denoted by the same letter with subscripts from 1 to s. The integrand will be denoted by

$$f(\underline{x}) = f(x_1,\ldots,x_s) .$$

About this function we shall assume that it has all its mixed partial derivatives up to

$$\frac{\partial^{rs} f}{\partial x_1^r \ldots x_s^r} \tag{1}$$

with $r \geq 1$ of bounded variation in the sense of Hardy and Krause over the closure $\bar{Q}^s$ of Q^s. Moreover, we shall assume that all the partial derivatives

$$\frac{\partial^{q(1)+\ldots+q(s)} f}{\partial x_1^{q(1)} \ldots \partial x_s^{q(s)}} \tag{2}$$

with $0 \leq q(i) \leq r-1$ $(i=1,\ldots,s)$ agree on opposite sides of $\bar{Q}^s$, that is, that substituting for any subset of $x_1,\ldots,x_s$ values 0 or 1 yields the same function. This, again, seems to be quite a serious restriction on the integrand. However, there are methods of reducing any integrand with the required degree of regularity to one which satisfies the last-mentioned condition (see, for instance, [2], or, for more explicit formulas, [6]).

It is well known that if a function $f(x)$ of one variable is of bounded variation over the interval $[0,1]$, then the Fourier coefficient C_h of the formal expansion

$$f(x) = \sum_{h-\infty}^{+\infty} C_h \exp(2\pi ihx) \tag{3}$$

is $O(h^{-1})$. This result can be extended to functions of several variables. The subscript h

becomes a vector $\underline{h} = \langle h_1, \ldots, h_s \rangle$, and it is convenient to introduce the notation

$$R(\underline{h}) = \prod_{j=1}^{s} \max(1, |h_j|) .$$

One proves, then, that if $f(\underline{x})$ is of bounded variation in the sense of Hardy and Krause over $\bar{Q}^s$, then [5]

$$|C_h| \leq K\, R(\underline{h})^{-1} ,$$

where K is the sum of the variations in the sense of Vitali in the appropriate numbers of dimensions of f and of all the functions obtained from it by substituting independently the values 0 or 1 in all the possible subsets of the set of coordinates $\langle x_1, \ldots, x_s \rangle$.

As an easy consequence one obtains the following:

<u>THEOREM 1</u>. If, for some $r > 0$, the partial derivatives of f up to (1) are of bounded variation in the sense of Hardy and Krause over $\bar{Q}^s$ and if the partial derivatives (2) with $0 \leq q(i) \leq r-1$ $(i=1,\ldots,s)$ agree on the opposite sides of $\bar{Q}^s$, then the Fourier coefficient in the formal expansion of f satisfies

$$|C_{\underline{h}}| \leq K^{(r)} R(\underline{h})^{-(r+1)} , \tag{4}$$

where $K^{(r)}$ is determined by the variations in the sense of Vitali in the appropriate numbers of dimensions of (1) and of all the functions obtained from it by substituting independently the values 0 or 1 in all the possible subsets of the set of coordinates.

Note that even with $r = 1$, this yields

$$|c_{\underline{h}}| = O(R(\underline{h})^{-2}).$$

Hence the Fourier series of f converges uniformly and absolutely, and it follows that it converges to f, and that it can be integrated term-by-term.

Now we can pass to good lattice points. We compute approximately the integral

$$\int_{Q^s} f(\underline{x})d\underline{x} \tag{5}$$

as follows: We choose an integer $m > 1$ and an s-dimensional lattice point $\underline{g}$ modulo m, that is a vector with integral coordinates, which we assume to be reduced modulo m. All the multiples of $\underline{g}$ reduced in each coordinate modulo m are obtained by multiplying $\underline{g}$ by $0,1,\ldots,m-1$. As an approximate value of (5) we take the sum

$$\frac{1}{m}\sum_{k=0}^{m-1} f\left(\frac{k}{m}\underline{g}\right).$$

In principle, the coordinates should be reduced, here, modulo 1, but if we think of f as being given by its Fourier expansion, this is no longer necessary, and the previous expression becomes

$$\frac{1}{m} \sum_{h \in L^s} C_h \sum_{k=0}^{m-1} \exp(2\pi i \, \frac{k}{m} \underline{g} \cdot \underline{h}) \ ,$$

where the dot denotes the scalar product, and L^s is the lattice formed in s dimensions by all the points having integers as coordinates.

When $\underline{g} \cdot \underline{h} \neq 0 (\mathrm{mod}\ m)$, the terms of the sum with respect to k form a cyclic geometric progression which differs from that of the m-th roots of 1 only by a constant. Thus such a sum is always equal to 0. If

$$\underline{g} \cdot \underline{h} \equiv 0 (\mathrm{mod}\ m), \tag{6}$$

all the exponentials are equal to 1, and the sum divided by m is equal to 1. On the other hand, clearly,

$$\int_{Q^s} f(\underline{x}) d\underline{x} = C_{\underline{0}} \ .$$

Hence the error of integration is exactly equal to the sum of all $C_{\underline{h}} (\underline{h} \in L^s)$ satisfying (6) and different from the zero vector, which will be denoted by $\underline{0}$. Thus, if f satisfies the conditions of Theorem 1, the absolute value of the error of integration is bounded by

$$K^{(r)} P^{(r+1)}(\underline{g}) \ ,$$

where

$$P^{(r+1)}(\underline{g}) = \sum R(\underline{h})^{-(r+1)},$$

the sum being taken over all lattice points $\underline{h} \neq \underline{0}$ satisfying (6).

It is convenient to denote by $\rho(\underline{g})$ the minimum of $R(\underline{h})$ for all lattice points $\underline{h} \neq \underline{0}$ satisfying (6). Given s and r, $P^{(r+1)}(\underline{g})$ is still not a function of $\rho(\underline{g})$; there are cases of different values of $P^{(r+1)}(\underline{g})$ corresponding to two vectors with the same $\rho(\underline{g})$. However, there are upper bounds for the former function which are, as one could expect, decreasing functions of $\rho(\underline{g})$ given s [6]. These bounds are rather complicated. Let it suffice to say here that for sufficiently large m,

$$P^{(2)}(\underline{g}) \leqslant \frac{2^{3s+1}(\log m)^{s-1}}{(s-1)!(\log 2)^{s-1}\rho(\underline{g})^2},$$

while, obviously, for any $r \geqslant 2$,

$$P^{(r+1)}(\underline{g}) < \rho(\underline{g})^{-(r-1)} P^{(2)}(\underline{g}) .$$

As a figure of merit of $\underline{g}$ one could take either $P^{(r+1)}(\underline{g})$ or $\rho(\underline{g})$, but the former function depends on r, which should be chosen according to the properties of the integrand. Hence there is much to be said for choosing $\rho(\underline{g})$ as the figure of merit, which we ought to minimize for a given m and s.

In the case of two dimensions, the practical problem is neatly solved [4], with the help of Fibonacci numbers, which are defined by

$$u_1 = u_2 = 1, \quad u_{n+2} = u_{n+1} + u_n \quad (n=1,2,\ldots) .$$

If we put $m = u_n$, $g_1 = 1$, $g_2 = u_{n-1}$, we have exactly $\rho(\underline{g}) = u_{n-2}$. The ratio $\rho(\underline{g})/m = u_{n-2}/u_n$ is known to tend to $\frac{1}{2}(3-\sqrt{5})$, when n, and, with it, $m = u_n$, tends to infinity. Apart from the trivial case of $m = u_4 = 3$, this is the best value of $\rho(\underline{g})/m$. From the practical viewpoint, this is about all we need to know when we integrate numerically a function over a square, but it must be admitted that integrating a reasonably smooth function in two dimensions is no great problem. Let it be noted, however, that we know very little about the case of an arbitrary m, even in two dimensions.

In more than two dimensions, no recipe is known for lattice points $\underline{g}$ yielding satisfactory values of $\rho(\underline{g})$, and there seems to be no reason to believe that such a recipe could be found. One has to find such lattice points by a laborious procedure of trial and error, or more efficiently, but still quite laboriously, by eliminating "bad" lattice points. There are, however, existence theorems which give us at least the assurance that the search for good lattice points will not be in vain.

Assume that the numbers m and s are given, and that we are looking for lattice points $\underline{g}$ yielding a value of $\rho(\underline{g})$ bigger than a chosen number ρ_0. We can eliminate those $\underline{g}$ for which $\rho(\underline{g}) \leq \rho_0$ in the following manner: We consider all $\underline{h} \in L^s$ with $R(\underline{h}) \leq \rho_0$; for each of them, we must eliminate all $\underline{g} \in L^s$ satisfying (6).

The first step consists of obtaining a good upper bound of the number of lattice points $\underline{h}$ under consideration, that is satisfying $R(\underline{h}) \leq \rho_0$. For any number t of dimensions, let $2N^{(t)}(\rho_0)$ be the number of lattice points $h^{(t)} = \langle h_1, \ldots, h_t \rangle$ for which $0 < |h_1 \ldots h_t| \leq \rho_0$. We have

$$N^{(t)}(\rho_0) =$$

$$\rho_0\left(\frac{(2 \log \rho_0)^{t-1}}{(t-1)!} + a_1^{(t)}(\log \rho_0)^{t-2} + \ldots + a_{t-1}^{(t)}\right),$$

where $|a_1^{(t)}|, \ldots, |a_{t-1}^{(t)}|$ have finite upper bounds depending only on t. One proves it by induction, noting that the formula is trivially correct for $t = 1$, and that we have the recurrence relation

$$N^{(t+1)}(\rho_0) =$$

$$2(N^{(t)}(\rho_0) + N^{(t)}(\rho_0/2) + \ldots + N^{(t)}(\rho_0/[\rho_0])).$$

If $2M^{(s)}(\rho_0)$ is the number of lattice points $\underline{h} \neq \underline{0}$ satisfying $R(\underline{h}) \leq \rho_0$, then obviously

$$M^{(s)} = \sum_{k=0}^{s-1} \binom{s}{k} N^{(s-k)}(\rho_0) .$$

Hence

$$M^{(s)}(\rho_o) = \rho_o((2 \log \rho_o)^{s-1}/(s-1)! + \alpha_1^{(s)}(\log \rho_o)^{s-2}+\ldots+\alpha_{s-1}^{(s)}),$$

where, again, $|\alpha_1^{(s)}|,\ldots,|\alpha_{s-1}^{(s)}|$ are bounded by numbers depending only on s.

It should be noted that, for any $h \in L^s$, the solutions of (6) coincide exactly with those of

$$(-\underline{h})\cdot\underline{g} \equiv 0(\text{mod } m) .$$

Consequently, when we eliminate "bad" lattice points, we need to take into account only half the number of lattice points $\underline{h} \neq 0$ with $R(\underline{h}) \leq \rho_o$, for instance those for which the nonzero coordinate with the highest subscript is positive. Thus all the "bad" lattice points are eliminated by $M^{(s)}(\rho_o)$, and <u>not</u> by $2M^{(s)}(\rho_o)$ congruences of the form of (6).

One more preliminary remark which does not depend on the nature of the integer m: If the latter is sufficiently large and

$$\rho_o = (s-1)!m/(2 \log m)^{s-1} ,$$

we have

$$M^{(s)}(\rho_o) < m .$$

Indeed, clearly $\log m - \log \rho_o \to \infty$ as $m \to \infty$, while

$$(7) \quad \frac{(2 \log m)^{s-1}}{(s-1)!} - \frac{(2 \log \rho_o)^{s-1}}{(s-1)!} >$$

$$> \frac{2^{s-1}(\log \rho_o)^{s-2}(\log m - \log \rho_o)}{(s-2)!}$$

Hence, for sufficiently large values of m,

(8) $$m = \rho_o \frac{(2 \log m)^{s-1}}{(s-1)!}$$

$$> \rho_o \left(\frac{(2 \log \rho_o)^{-1}}{(s-1)!} + \alpha_1^{(s)} (\log \rho_o)^{s-2} + \ldots + \alpha_{s-1}^{(s)} \right)$$

$$= M^{(s)}(\rho_o).$$

It should be noted that the last inequality does not depend on the particular nature of $\alpha_1^{(s)}, \ldots, \alpha_{s-1}^{(s)}$, provided that these coefficients should be bounded by a number depending only on s.

The congruence (6) is obviously unaffected if we multiply $\underline{g}$ by any integer relatively prime to m. Hence, when m is a prime, we can assume $g_1 = 1$, which is convenient in many ways. Then (6) can be written as

(9) $$h_1 + h_2 g_2 + \ldots + h_s g_s \equiv 0 \pmod m .$$

We assume here $\underline{h} \neq \underline{0}$. With the exception of the case when $h_1 \neq 0$, $h_2 = h_3 = \ldots = h_s = 0$, such a congruence is satisfied by m^{s-2} lattice points of the type under consideration, which are eliminated in our search for good lattice points. According to (8), the number of lattice points eliminated in this manner is smaller than $M^{(s)}(\rho_o)m^{s-2} < m^{s-1}$. Since there are in all m^{s-1} lattice points modulo m with $g_1 = 1$, we are still left with some lattice points of the type under consideration. For these, (9) clearly entails $R(\underline{h}) > \rho_0$. Hence we have proved the following existence theorem:

THEOREM 2. For any number s of dimensions, and for any prime m exceeding a number which depends only on s, there exist lattice points modulo m satisfying

$$\rho(\underline{g}) > \frac{(s-1)!\, m}{(2 \log m)^{s-1}} .$$

We shall describe such lattice points as good lattice points.

The information contained in this theorem is very incomplete. First of all, the assumption that m should be a prime leads to a remarkably simple proof, but there are excellent reasons to believe that it is by no means necessary for the validity of the theorem. The case of two dimensions illustrates it in a striking manner, since we obtain the best lattice points when m is a Fibonacci number. But the Fibonacci number u_n with $n > 4$ can be a prime only if n is a prime, and even then it may be composite. As a matter of fact, it is not even known whether there are infinitely many primes among the Fibonacci numbers.

Tables of good lattice points were obtained in 3 and 4 dimensions by Dominique Maisonneuve [3]. For easily understandable reasons, these tables were confined to good values of m, which incidentally, greatly simplified their compilation. We call a value of m good in the context of a given number of dimensions if a lattice point $\underline{g}$ modulo this value yields a value of $\rho(\underline{g})$ bigger than any which can be obtained with a smaller value of m.

In three dimensions, m goes from 21 with $\rho(\underline{g}) = 3$ to 1958 with $\rho(\underline{g}) = 162$. In four dimensions, m goes from 52 with $\rho(\underline{g}) = 3$ to 1354 with $\rho(\underline{g}) = 40$. These tables will be pushed further, but even in their present state they allow as to compute integrals, even in four dimensions, with quite a high degree of accuracy and in a very short time. Extensions to more than four dimensions are also planned. Among the good values of m obtained in this way, primes are rare exceptions. Partial results in the theory of good lattice points modulo composite numbers will be discussed further on in this article.

The numerical tables in question also show that the existence theorem does not take into account the difference between good and bad values of m. Apart from the case of two dimensions, we have little evidence concerning "bad" values of m, but in two dimensions one notices, for instance, that with $m = 34 = u_9$ of Fibonacci and $\underline{g} = \langle 1, u_8 \rangle$, we have $\rho(\underline{g}) = u_7 = 13$, while with $m = 35$, the best possible value of $\rho(\underline{g})$ is 8, attainable with $m = 21 = u_8$.

The lower bound of the best value of $\rho(\underline{g})$ given in Theorem 2 is certainly much too low, at least for good values of m. For three dimensions, Table 1 gives the values of the best $\rho(\underline{g})$ and of the corresponding ratios $\rho(\underline{g})\log m/m$ for all the good lattice points obtained, while for four dimensions, Table 2 gives the best values of $\rho(\underline{g})$ and the corresponding ratios $\rho(\underline{g})(\log m)^2/m$. Looking at these tables, one notes immediately that the oscillations of these values cannot obscure the increasing trend of the ratios. In either case, the smallest value of the ratio corresponds to the smallest good m, while the biggest value of the ratio corresponds in three dimensions to the last-but-one value of m, and in four dimensions to the last, that is, the biggest value of m. The numerical data gathered in tables 1 and 2 support the conjecture that in three dimensions for sufficiently large good values of m there exist lattice points yielding

$$\rho(\underline{g}) > m/(2 \log m) ,$$

and that in four dimensions for sufficiently large good values of m (apparently from $m = 60$ onwards) there exist lattice points yielding

$$\rho(\underline{g}) > m/(\log m)^2$$

The much higher values of $\rho(\underline{g})$ than guaranteed by Theorem 2 are easily explained by the fact that many "bad" lattice points are eliminated by more than one congruence of the type of (6), so that the total number of eliminated lattice points is much smaller than m^{s-2} multiplied by the number of such congruences. Unfortunately, the evaluation of the average number of congruences (6) with $R(\underline{h}) \leq \rho_o$ which eliminate a "bad" lattice point $\underline{g}$ seems to be a very difficult problem.

The proof that at least such a weak theorem as Theorem 2 is valid for arbitrary composite moduli m will certainly be far from simple. However, here are some preliminary considerations and very incomplete results obtained by still simple and elementary arguments. In comparison with the case of a prime modulus, two effects have to be taken into consideration:

(i) Now a congruence of the type of (9) no longer has exactly m^{s-2} solutions in $\underline{g}$ modulo m.

(ii) The congruences

(10) $h_j g_j \equiv 0 (\mathrm{mod}\ m) \ (|h_j| \leq \rho_o;\ j = 1,\ldots,s)$,

which, with a prime m, eliminated only $g_j = 0$, now eliminate all the values of g_j which have with m a common divisor bigger than, or equal to, m/ρ_o.

These two effects will be discussed separately, beginning with the first. The following lemma is proved without any difficulty.

<u>LEMMA</u>. If $(m,h_2,\ldots,h_s) = 1$, then (9) has exactly m^{s-2} solutions modulo m in $\underline{g}$.

An immediate consequence of it is that if

$$(m,h_1,h_2,\ldots,h_s) = (m,h_2,\ldots,h_s) = D,$$

then (9) has exactly $m^{s-2}D$ solutions in $\underline{g}$ modulo m. On the other hand, if

$$(m,h_2,\ldots,h_s) > (m,h_1,\ldots,h_s) ,$$

then (9) has no solutions in $\underline{g}$.

It follows that the upper bound in terms of m and s for the total number of solutions modulo m in $\underline{g}$ of all the congruences of the type of (9) with $R(\underline{h}) \leqslant \rho_o$ and $h_1 \neq 0$ which was found in the case of a prime modulus m is still valid when m is composite. Indeed, if

$$(m,h_2,\ldots,h_s) = D > 0,\ h_1 = h_1'D,\ h_1' > 0 ,$$

we can associate with the given congruence a set of $D-1$ congruences without solutions in $\underline{g}$, namely those which are obtained from it by substituting for h_1 in turn

$$(h_1'-1)D+1,\ldots,(h_1'-1)D+D-1 .$$

Obviously, the case of $h_1 < 0$ is similar, and the association is one-to-one. We note, incidentally, that outside these classes there still remain congruences without solutions. Indeed, if $R(\underline{h}) \leqslant \rho_o$, $(m,h_2,\ldots,h_s) = D$, $h_2\ldots h_s D > \rho_o$, and $h_1 \neq 0$, then $0 < |h_1| < D$, and consequently the congruence has no solution in $\underline{g}$; however, it corresponds, in the sense explained above, to no congruence with $(m,h_1,h_2,\ldots,h_s) = D$.

The whole argument does not apply when $h_1 = 0$. Since we treat separately congruences of the type of (10), it deals nevertheless completely with the case of $s = 2$.

Now let $s \geq 3$ and $2 \leq t \leq s-1$. We consider the case when $1 < i(1) < \ldots < i(t) \leq s$ and $h_{i(1)}\ldots h_{i(t)} \neq 0$, the remaining $s - t$ coordinates of h being all equal to 0. If

$$(m,h_{i(1)},\ldots,h_{i(t)}) = D > 1 ,$$

the congruence has exactly $m^{s-2}D$ solutions in $\underline{g}$. For a given D and t, the number of such congruences is, clearly

$$\binom{s-1}{t} N^{(t)}(\rho_o/D^t) < \binom{s-1}{t} N^{(t)}(\rho_o)D^{-t} ,$$

bringing the upper bound for the number of solutions of such congruences to

$$\binom{s-1}{t} N^{(t)}(\rho_o)\, D^{1-t} m^{s-2} . \tag{11}$$

We have to consider the various values which D can take within the interval $[1,[\rho_o^{1/t}]]$. In the absence of a good upper bound for the number of such divisors of m, as a very rough upper bound for the sum of (11), we can take

$$(12) \qquad \binom{s-1}{t} N^{(t)}(\rho_o)\ m^{s-2} \sum_{D=1}^{[\rho_o^{1/t}]} D^{1-t} \ .$$

If $t > 2$, the sum at the end of the last expression is bounded, so that the whole expression is $m^{s-2}O(\rho_o(\log \rho_o)^{t-1})$, and since $t \leqslant s-1$, the coefficient of m^{s-2} is absorbed by the left-hand side of (7). If, on the other hand, $t = 2$, then $\Sigma D^{1-t} = O(\log \rho_o)$, and (12) is of the order of $m^{s-2}\rho_o(\log \rho_o)^2$. If $s \geqslant 4$, this term is still absorbed by the left-hand-side of (7). Thus whenever $s \geqslant 4$, the additional number of eliminations of bad lattice points is of a lower order of magnitude than the left-hand-side of (7), so that, for a sufficiently large m, the total number of eliminations (not counting those arising out of (10))is still smaller than m^{s-1}, like in the case of a prime modulus.

The case of $s = 3$ is more complicated. However, a very simple argument shows that at least if the total number of divisors of m not exceeding ρ_o is bounded by $C \log m$, where C is a constant, the number of additional solutions of (9) in $\underline{g}$ which are due to congruences

$$h_2 g_2 + h_3 g_3 \equiv 0 (\text{mod } m)$$

with $0 < |h_2 h_3| < \rho_o$ and $(m, h_2, h_3) > 1$ is still absorbed by the left-hand side of (7). Indeed, having to take the sum of (11) for all the divisors of m, we note that the summand is a decreasing function of D. Hence the sum in question is smaller than

$$(13) \qquad \binom{s-1}{t} N^{(t)}(\rho_o) m \sum_{D=1}^{[C \log m]} D^{-1}$$

$$\sim 2\rho_o m \log \rho_o \operatorname{loglog} m \ .$$

On the other hand, the right-hand side of (7) becomes

$$2^2 \log \rho_o (\log 2 + 2 \operatorname{loglog} m)$$

$$> 8 \log \rho_o \operatorname{loglog} m \ .$$

Hence if we add (13) to the number of eliminations of bad values of $\underline{g}$, we still find, for sufficiently large values of m, and with the same value of ρ_o in terms of m as previously,

$$m^2 > mM^{(s)}(\rho_o) + m\, \rho_o \operatorname{loglog} m \ .$$

In other terms, the number of eliminations by congruences other than (10) is still smaller than the total number of possible values of $\underline{g}$ modulo m, namely m^2.

There remain to be considered the congruences of the type of (10). It is clear that here we can confine ourselves to the case where h_j is what we should like to describe as a ρ_o-maximum divisor of m. By this we mean a divisor of m which is smaller than, or equal to, ρ_o, but divides no other divisor of m having this property. Heuristic arguments which will not be reproduced here indicate that among the lattice points eliminated by other congruences, the proportion of those satisfying at the same time at least one congruence of the type of (10) is such that the additional number of eliminations of bad lattice points due to the latter type of congruence is asymptotically negligible when $m \to \infty$. A formalization of these arguments, necessarily quite involved, would yield a proof of Theorem 2 without the assumption of m being a prime, at least when $s \neq 3$. For the time being, without such intricate considerations, we can easily prove the following results:

PROPOSITION 3. Let $s \neq 3$. If, with

$$\rho_o = (s-1)!m/(2 \log m)^{s-1} ,$$

for a class of values of m the number of ρ_o-maximum divisors is bounded by $C(\log \rho_o)^{s-2}$, where C is a constant, then there exist lattice points $\underline{g}$ with $\rho(\underline{g}) > \rho_o$ for every sufficiently large m in this class.

Indeed, any ρ_o-maximum divisor of m eliminates at most ρ_o values of any $g_j (j=1,...,s)$, thus fewer than $(s-1)\rho_o m^{s-2}$ lattice points, the total being of a smaller order of magnitude than the left-hand side of (7). Together with the previously discussed number of additional eliminations due to congruences with $(m,h_1,...,h_s) > 1$, this is still a number of a smaller order of magnitude than the left-hand side of (7), so that, when m is sufficiently large, the number of eliminations is still smaller than the number of different lattice points modulo m.

Remark. The condition on m as stated above, is a little cumbersome. It may be rather more convenient to replace the bound on the number of ρ_o-maximum divisors of m by $C(\log m)^{s-2}$ which is legitimate, since clearly, $\log m/\log \rho_o \to 1$ as $m \to \infty$. Given C and s, there is no difficulty in finding an explicit formula for a number μ such that the conclusion of the preceding proposition applies to all m bigger than μ in the class under consideration, but this is of little practical importance. Obviously, the condition on m is satisfied if the total number of divisors of m is bounded by $C(\log m)^{s-2}$.

It is well known that the total number of divisors of any integer n is $O(n^\delta)$ for any positive δ. Hence the last condition on m will be satisfied, in particular, whenever, in the class under consideration:

(a) m has a prime divisor bigger than, or equal to, $[\rho_o]$, or:

(b) m has two prime divisors, say p_1 and p_2, satisfying $p_1 p_2 \geq \rho_o$ and $p_i \leq [\rho_o]^2/m (i = 1,2)$.

Actually, in either of these cases, the number of ρ_o-maximum divisors of m has a finite bound. Indeed, in the first case, if the prime divisor in question is p, and if $m = pm'$, then the only ρ_o-maximum divisors of m are m' and, if $p = [\rho_o]$, also p. In the second case, if $m = p_1p_2m'$, we have $m' \leqslant m/[\rho_o]$, so that $p_im' \leqslant [\rho_o]$ $(i = 1,2)$. Consequently, the only ρ_o-maximum divisors of m are p_1m', p_2m' and, if $p_1p_2 = [\rho_o]$, also p_1p_2.

Together with the previously obtained results concerning the additional number of eliminations due to the congruences of the type of (9) when $(m,h_1,h_2,h_3) > 1$ with $s = 3$, an easy adaptation of the proof of Proposition 3 yields:

<u>PROPOSITION 4</u>. Let $s = 3$. If, with ρ_o given by the previous formula, which now reduces to

$$\rho_o = \frac{1}{2}m(\log m)^{-2} ,$$

for a class of values of m the total number of divisors is bounded by $\log \rho_o$ multiplied by a constant, there exist lattice points $\underline{g}$ with $\rho(\underline{g}) > \rho_o$ for every sufficiently large m of this class.

Good lattice points modulo composite numbers were already considered by Korobov [2] in more than two dimensions. However, he confined himself to the case of products of two primes, and his promised figure of merit was less favorable than in the case of a prime modulus. It may, therefore, be worth stressing that in all the cases considered here the figure of merit, in our case $\rho(\mathbf{g})$, was given exactly by the same formula which was obtained in the case of prime moduli. The theorem which follows, rather weak in itself, though apparently capable of useful applications, shows that, on the contrary, there are cases where a composite modulus appears to be slightly more favorable than a prime one, insomuch as the lower bound for the best $\rho(\underline{g})$ which we are able to derive as a function of m and s is bigger than the expression obtained in Theorem 2 for the case of a prime modulus. It is a moderately generalized version of a proposition proved in [6].

PROPOSITION 5. Let the number $s (s \geq 3)$ of dimensions and the constant C be fixed. We consider a class of moduli m which are capable of being factorized into $m = m'm''$ with

$$m'' \leq \frac{2 \log m'}{s-1}, \tag{14}$$

there being not more than $C(\log m)^{s-3}$ numbers smaller than m' and not relatively prime to it. Then to each sufficiently large number m of this class there corresponds a lattice point $\underline{g}$ modulo m with

$$\rho(\underline{g}) > \frac{(s-1)!m}{(2 \log m')^{s-1}}. \tag{15}$$

The advantage consists here of having $\log m'$ instead of $\log m$ in the denominator of the right-hand side of the last inequality.

Proof. According to Proposition 3 when $s \neq 4$, and to Proposition 4 when $s = 4$, there exists a lattice point $\underline{g}^* = \langle g_2^*, \ldots, g_s^* \rangle$ modulo m' in $s-1$ dimensions satisfying

$$\rho(\underline{g}^*) > \frac{(s-2)!\ m'}{(2 \log m')^{s-2}} ,$$

We can, then, form in s dimensions a lattice point modulo m' of the form

$$\tilde{\underline{g}} = \langle 1, \lambda g_2^*, \ldots, \lambda g_s^* \rangle ,$$

choosing λ so that

$$(16) \qquad \rho(\tilde{\underline{g}}) > \frac{(s-1)!\ m'}{(2 \log m')^{s-1}} = \rho^* .$$

Indeed, among the m' possible values of λ we have to eliminate only those which satisfy

$$h_1 + \lambda(h_2^* g_2 + \ldots + h_s g_s^*) \equiv 0 (\text{mod } m')$$

with $\underline{h} \neq 0$ and $R(\underline{h}) \leqslant \rho^*$.

When $h_1 \neq 0$, we deal with the case of

$$(m', h_2 g_2^* + \ldots + h_s g_s^*) > 1$$

in the same way as in the proofs of Propositions 3 and 4. If $h_1 = 0$, we only have to exclude the values of $\lambda (1 \leqslant \lambda \leqslant m)$ which are not prime relative to m' because, when $(\lambda, m') = 1$, the congruences

$$\lambda(h_2 g_2^* + \ldots + h_s g_s^*) \equiv 0 (\text{mod } m')$$

and

$$(h_2 g_2^* + \ldots + h_s g_s^*) \equiv 0 (\text{mod } m')$$

are equivalent. Thus the number of values of λ which are eliminated is at most

$$M^{(s)}(\rho^*) + C(\log m')^{s-3} .$$

But we have seen, before proving Theorem 2, that this number is smaller than m' whenever m' is sufficiently large, the term $C(\log m')^{s-3}$ being again absorbed by the left-hand side of (7). Hence there remain some of the m' possible values of λ which are not excluded, and, therefore, satisfy (16).

Now we form the lattice point

$$\underline{g} = \langle 1, m''\lambda g_2^*, \ldots, m''\lambda g_s^* \rangle \text{ modulo } m .$$

It turns out that $\underline{g}$ satisfies (15). Indeed, if (6) is satisfied and $\underline{h} \neq 0$, but $h_1 = 0$, then the congruence (6) reduces to

$$g_2^* h_2 + \ldots + g_s^* h_s \equiv 0 (\text{mod } m') ,$$

and consequently, owing to (14),

$$R(\underline{h}) > \frac{(s-2)!\ m'}{(2 \log m')^{s-2}} = \frac{(s-1)!\ m'}{(2 \log m')^{s-1}} \cdot \frac{2 \log m'}{s-1}$$

$$\geqslant \frac{(s-1)!\ m}{(2 \log m')^{s-1}} .$$

If (6) is satisfied with $h_1 \neq 0$, m'' must divide h_1. Let $h_1 = h_1' m''$. Then the congruence reduces to

$$h_1' + \lambda g_2^* h_2 + \ldots + \lambda g_s^* h_s \equiv 0 (\text{mod } m') .$$

But then, according to (16),

$$R(\langle h_1', h_2, \ldots, h_s \rangle) > \rho^* ,$$

and consequently,

$$R(\underline{h}) = m'' R(\langle h_1', h_2, \ldots, h_s \rangle) > m'' \rho^* = \frac{(s-1)!\ m}{(2 \log m')^{s-1}} .$$

Thus (15) is satisfied, and this completes the proof.

It should be noted that the previous argument could also be applied to a better lower bound for $\rho(\underline{g})$ than that which was established in Propositions 3 and 4. In general, the procedure outlined here can be applied to obtain a good lattice point in s dimensions $(s \geq 3)$ whenever one is given in $s-1$ dimensions. Of course, this procedure is not guaranteed to yield always the best possible lattice points, but our tables of good lattice points [3] show that it is often the case. In three dimensions, out of 82 good lattice points, 31 could have been obtained by this method with $m'' = 2$, and 5 with $m'' = 3$; in four dimensions, out of 34 good lattice points, 16 could have been obtained by this method with $m'' = 2$.

The author regrets to have presented such an incomplete picture of the theory of good lattice points, but hopes to be forgiven because our knowledge of these problems is still rudimentary.

REFERENCES

[1] E. Hlawka, Zur angenäherten Berechnung mehrfacher Integrale, Monatsh. Math. 66 (1962), 140-151.

[2] N.M. Korobov, Teoretikocislovye metody v priblizhennom analize (Numbertheoretical methods in approximate analysis), Fizmatgiz, Moscow, 1963.

[3] Dominique Maisonneuve, Recherche et utilisation des "bons treillis": Programmation et résultats numériques in Applications of Num-Theory to Numerical Analysis, Proceedings of the Symposium at the Centre for Research in Mathematics, University of Montreal, September 9-14, 1971, Edited by S.K. Zaremba, Academic Press, Inc., New York, 1972.

[4] S.K. Zaremba, Good lattice points, discrepancy, and numerical integration, Ann. Mat. pura appl. (iv) 73 (1966), 293-318.

[5] S.K. Zaremba, Some applications of multidimensional integration by parts, Ann. Polon. Math. 21 (1968), 85-96.

[6] S.K. Zaremba, La méthode des "bons treillis" pour le calcul des integrales multiples in Applications of Number Theory to Numerical Analysis, Proceedings of the Symposium at the Centre for Research in Mathematics, University of Montreal, September 9-14, 1971, Edited by S.K. Zaremba, Academic Press, Inc., New York, 1972.

Note:

After the Symposium, the author proved that the existence theorem for good lattice points (Theorem 2 in the article) is still valid when the modulus m is a completely arbitrary natural number. The proof, which is no longer elementary, will be published in due course.

Centre de Recherches Mathématiques, Université de Montreal, Montreal, Quebec, Canada

Table 1: s = 3

m	$\rho(\underline{g})$	$\log m\ \rho(\underline{g})/m$
21	3	0.43493
26	4	0.50125
35	5	0.50791
44	6	0.51603
66	8	0.50784
78	9	0.50270
86	10	0.51795
98	12	0.56142
135	13	0.47236
142	14	0.48860
148	15	0.50647
150	16	0.53447
172	18	0.53869
185	20	0.56436
214	21	0.52657
222	24	0.58407
266	27	0.56675
284	28	0.55694
298	30	0.57353
301	33	0.62570
366	36	0.58059
418	40	0.57756
436	45	0.62728
504	46	0.56793
539	48	0.56012
546	52	0.60025

Table 1 (continued)

m	ρ(g)	log m ρ(g)/m
588	54	0.58562
597	55	0.58837
650	57	0.56798
706	58	0.53889
712	63	0.58116
737	64	0.57336
755	65	0.57051
759	66	0.57670
813	70	0.57694
828	72	0.58426
882	76	0.58441
909	81	0.60704
962	84	0.59979
1010	86	0.58903
1046	90	0.59823
1086	96	0.61792
1199	105	0.62083
1220	108	0.62911
1365	111	0.58703
1382	112	0.58604
1459	114	0.56926
1466	117	0.58183
1504	120	0.58371
1546	126	0.59849
1602	128	0.58958
1611	132	0.60507

Table 1 (continued)

m	$\rho(\underline{g})$	$\log m\ \rho(\underline{g})/m$
1626	140	0.63662
1680	144	0.63656
1764	146	0.61871
1838	160	0.65431
1958	162	0.62712

Table 2: $s = 4$

m	$\rho(\underline{g})$	$(\log m)^2\rho(\underline{g})/m$
52	3	0.90071
60	4	1.11758
97	5	1.07876
118	6	1.15726
168	7	1.09396
180	8	1.19852
222	9	1.18333
266	10	1.17201
286	12	1.34225
402	14	1.25225
440	15	1.26303
445	16	1.33704
541	18	1.31780
562	20	1.42662
718	22	1.32521
732	24	1.42637
930	25	1.25591
932	26	1.30416
964	30	1.46925
1142	32	1.38898
1234	36	1.47810
1354	40	1.53607